円周率 100,000,000 桁表
縮刷版

- 1行1,000桁、1ページ500行500,000桁が掲載されています。

- 正確な値になるように十分注意を払いましたが、億が一、掲載した値が間違っていたとしても、発行者は責任をとれません。

- 落丁、乱丁は（在庫がある限り）お取り替えします。

取扱い上の注意事項

1. 本書の主要部分は、精細な印刷がなされております。
2. 印刷特性上のカスレが存在しますが、これは仕様です。
3. 紙面に対して消しゴムをかけたり強くこすったり等すると、印刷の品質が損なわれる恐れがあります。

円周率 100,000,000 桁表　　　　　　　　　　　00000001–00500000

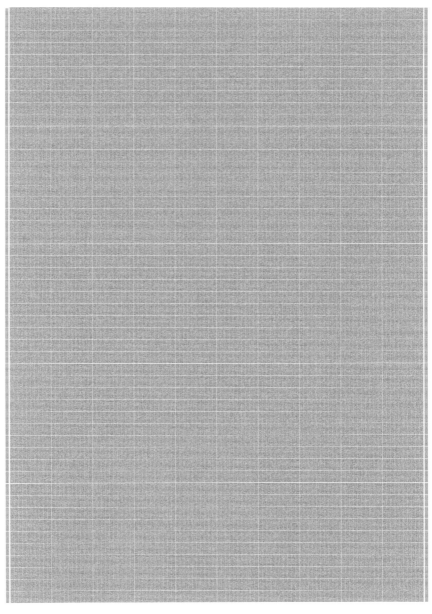

π upto 100,000,000 decimal digits　　　　　00000001–00500000

00500001–01000000 円周率 100,000,000 桁表

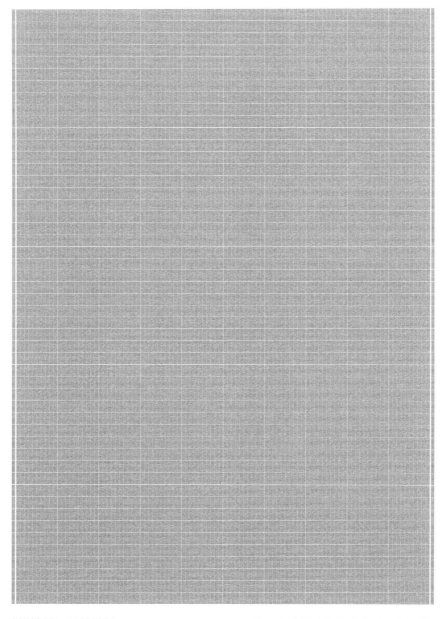

00500001–01000000 π upto 100,000,000 decimal digits

円周率 100,000,000 桁表　　　　　　　　　01000001–01500000

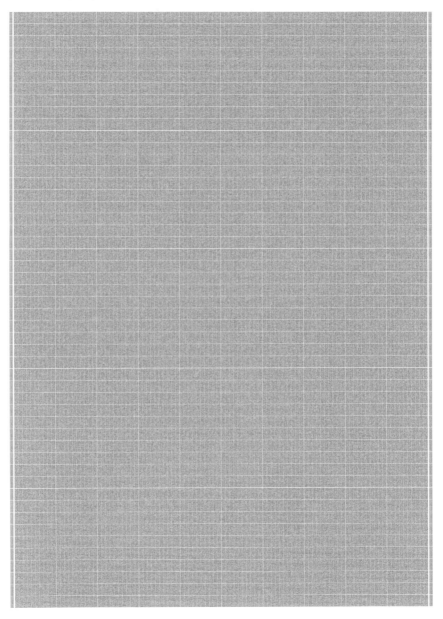

π upto 100,000,000 decimal digits　　　　01000001–01500000

01500001–02000000 円周率 100,000,000 桁表

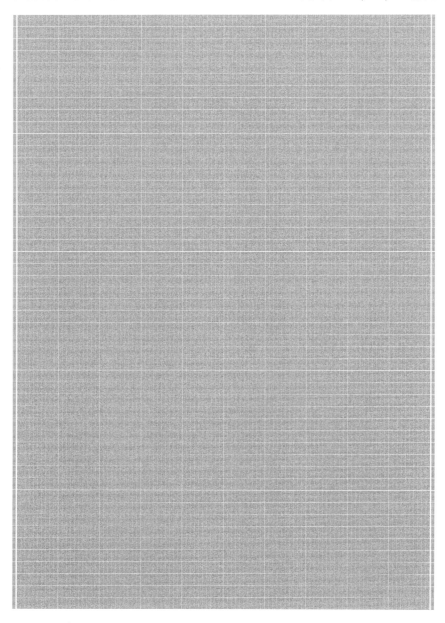

01500001–02000000 π upto 100,000,000 decimal digits

円周率 100,000,000 桁表　　　02000001–02500000

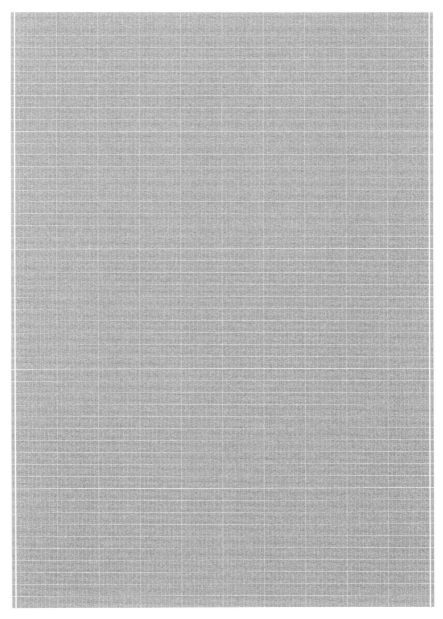

π upto 100,000,000 decimal digits　　　02000001–02500000

02500001–03000000 円周率 100,000,000 桁表

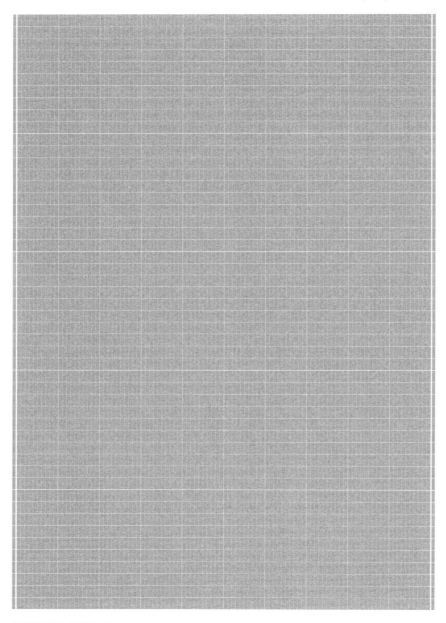

02500001–03000000 π upto 100,000,000 decimal digits

円周率 100,000,000 桁表 03000001–03500000

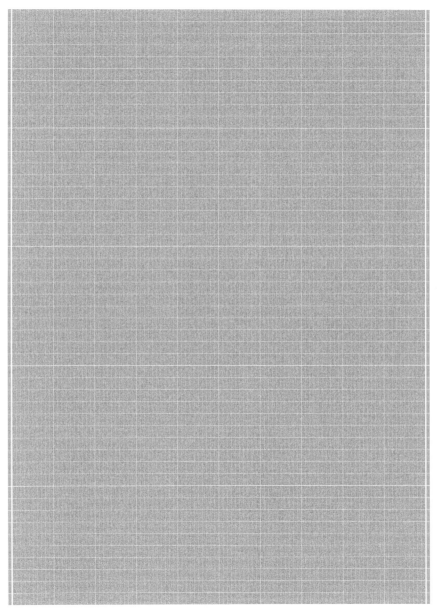

π upto 100,000,000 decimal digits 03000001–03500000

03500001–04000000 円周率 100,000,000 桁表

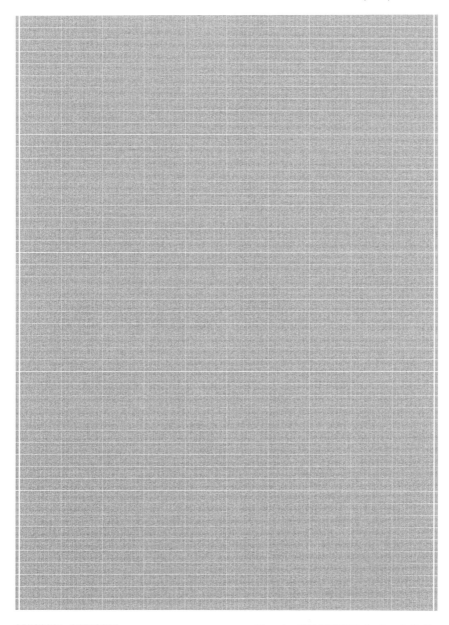

03500001–04000000 π upto 100,000,000 decimal digits

円周率 100,000,000 桁表　　　　　　　　　　04000001–04500000

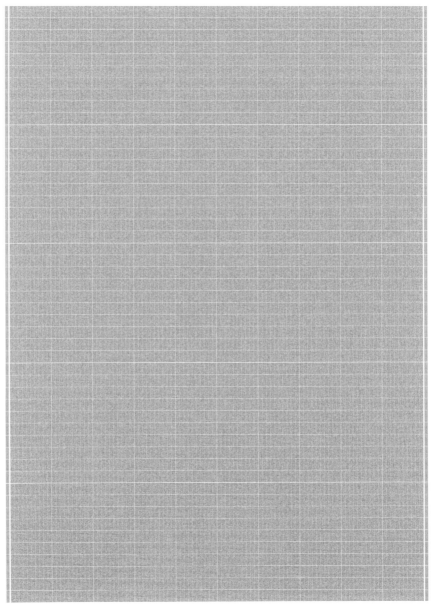

π upto 100,000,000 decimal digits　　　　　04000001–04500000

04500001–05000000 円周率 100,000,000 桁表

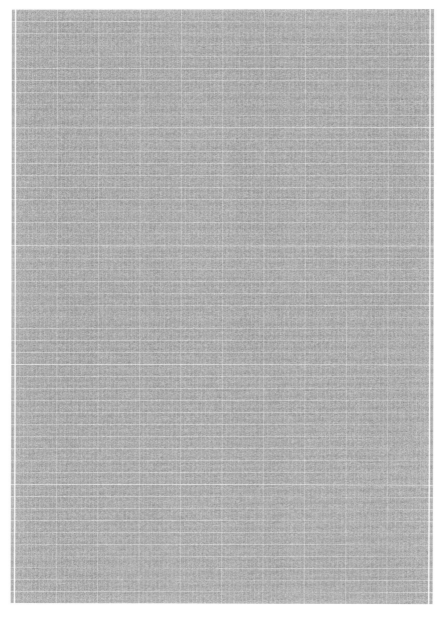

04500001–05000000 π upto 100,000,000 decimal digits

円周率 100,000,000 桁表　　　　05000001–05500000

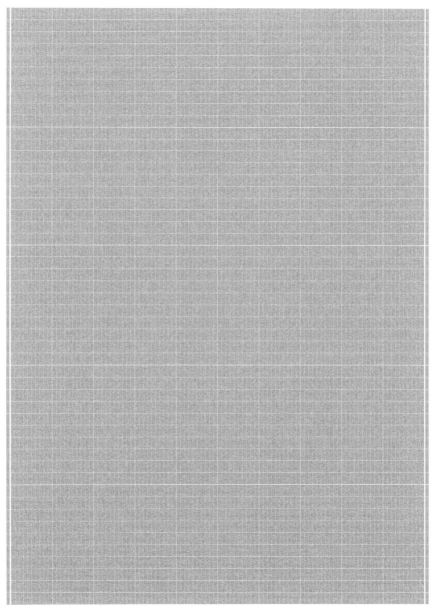

π upto 100,000,000 decimal digits　　　　05000001–05500000

05500001–06000000 円周率 100,000,000 桁表

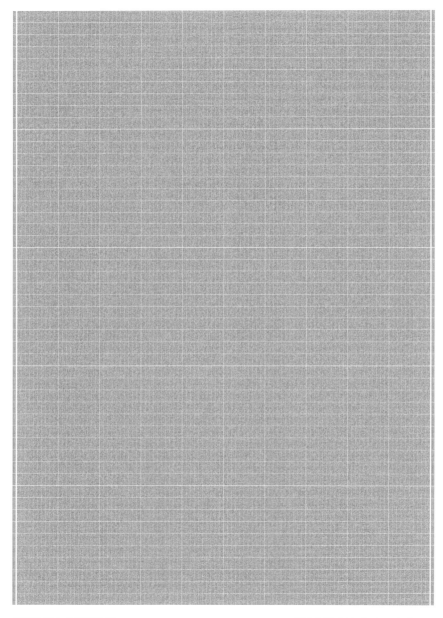

05500001–06000000 π upto 100,000,000 decimal digits

円周率 100,000,000 桁表　　　　　　　　　　06000001–06500000

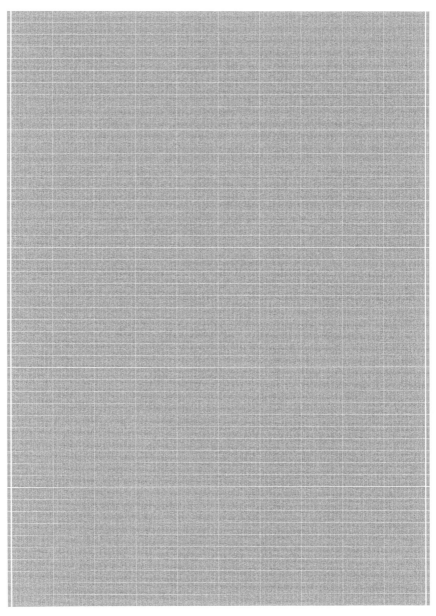

π upto 100,000,000 decimal digits　　　　　　06000001–06500000

06500001–07000000 円周率 100,000,000 桁表

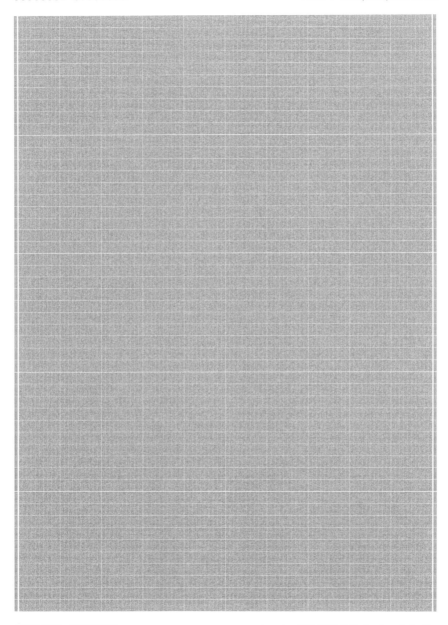

06500001–07000000 π upto 100,000,000 decimal digits

円周率 100,000,000 桁表 07000001–07500000

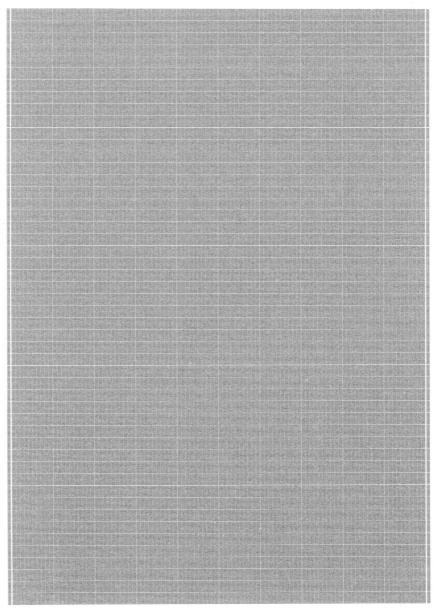

π upto 100,000,000 decimal digits 07000001–07500000

07500001–08000000 円周率 100,000,000 桁表

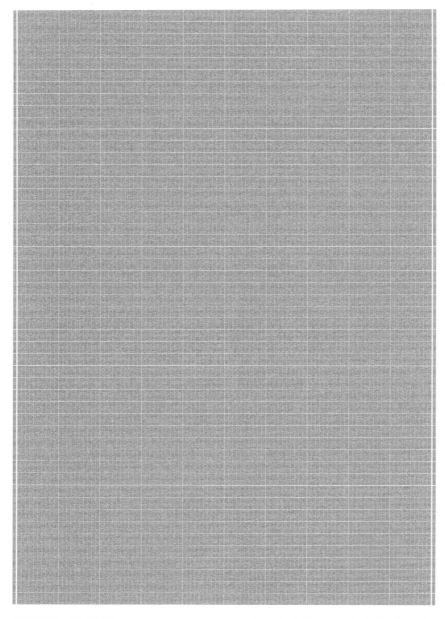

07500001–08000000 π upto 100,000,000 decimal digits

円周率 100,000,000 桁表　　　　　　　　　　08000001–08500000

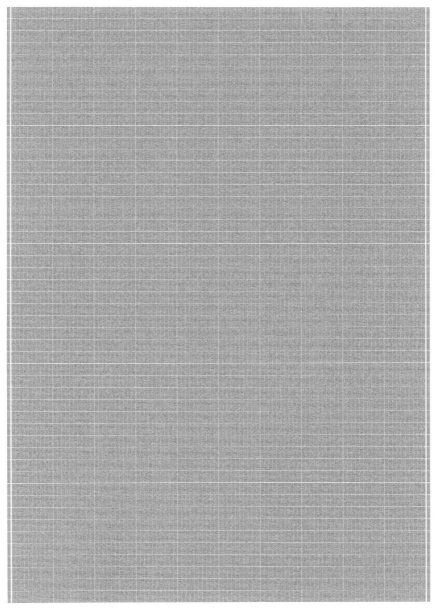

π upto 100,000,000 decimal digits　　　　　08000001–08500000

08500001–09000000 円周率 100,000,000 桁表

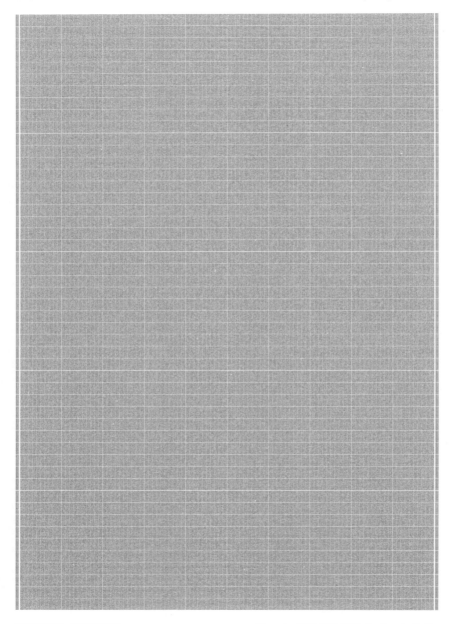

08500001–09000000 π upto 100,000,000 decimal digits

円周率 100,000,000 桁表　　　　　　　　　09000001–09500000

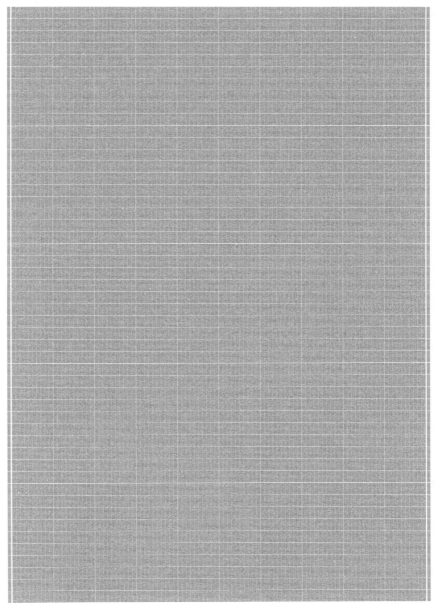

π upto 100,000,000 decimal digits　　　　　09000001–09500000

09500001–10000000 円周率 100,000,000 桁表

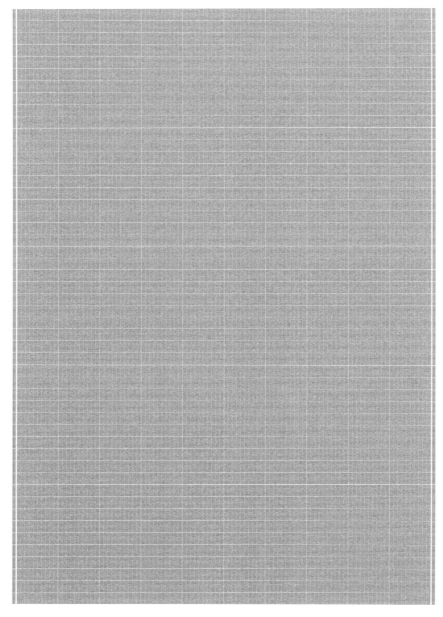

09500001–10000000 π upto 100,000,000 decimal digits

円周率 100,000,000 桁表　　　　　　　　　　10000001–10500000

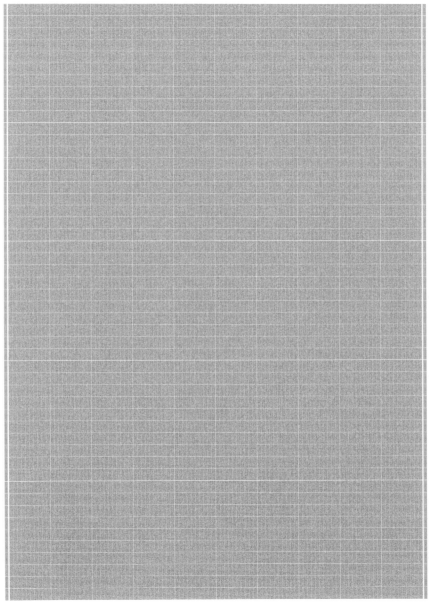

π upto 100,000,000 decimal digits　　　　　　10000001–10500000

10500001–11000000 円周率 100,000,000 桁表

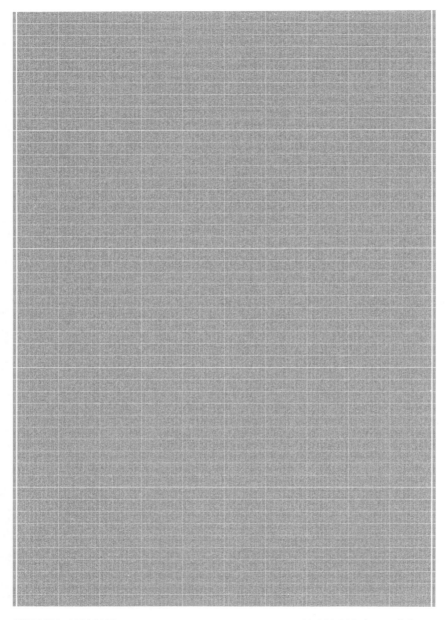

10500001–11000000 π upto 100,000,000 decimal digits

円周率 100,000,000 桁表 11000001–11500000

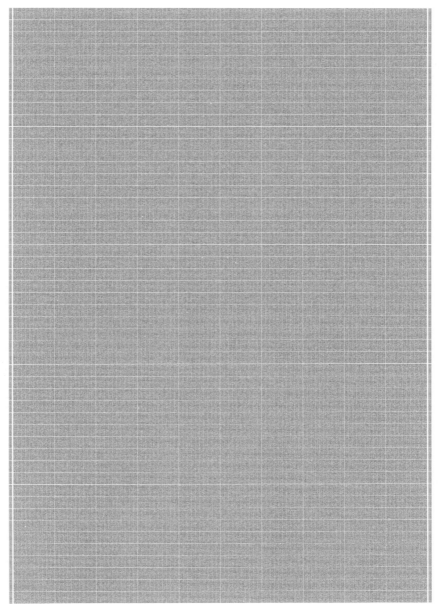

π upto 100,000,000 decimal digits 11000001–11500000

11500001–12000000 円周率 100,000,000 桁表

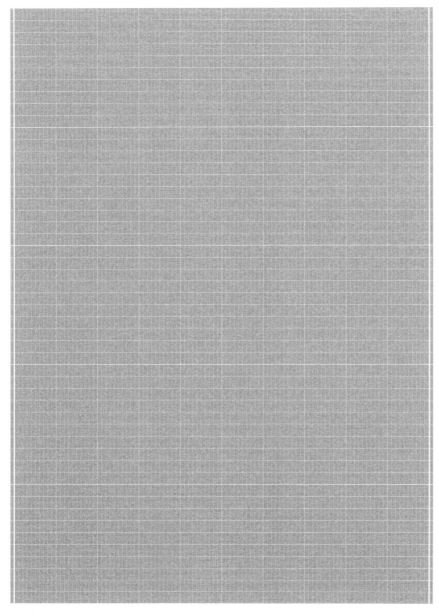

11500001–12000000 π upto 100,000,000 decimal digits

円周率 100,000,000 桁表　　　　　　　　12000001–12500000

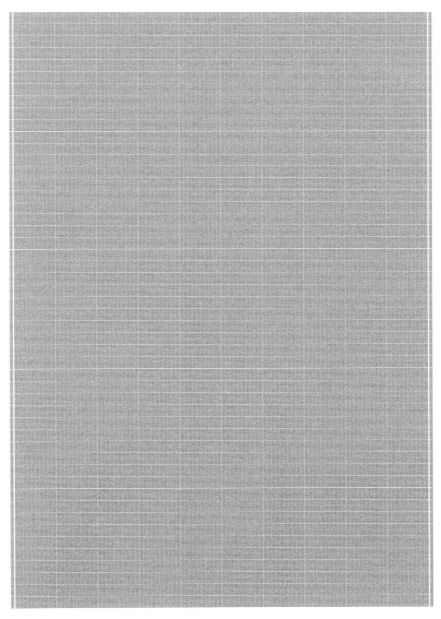

π upto 100,000,000 decimal digits　　　　12000001–12500000

12500001–13000000 円周率 100,000,000 桁表

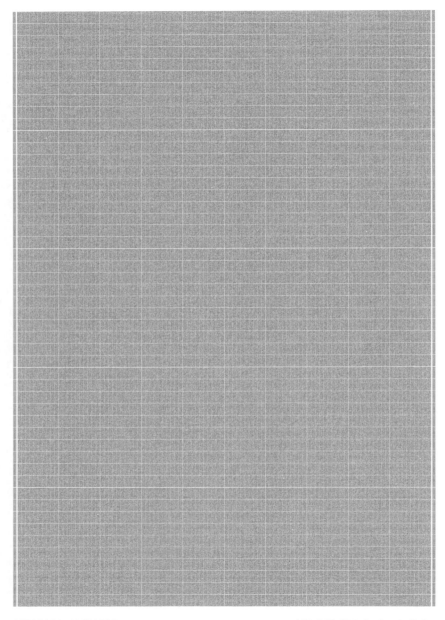

12500001–13000000　　　　　　　π upto 100,000,000 decimal digits

円周率 100,000,000 桁表　　　　　　　　　13000001–13500000

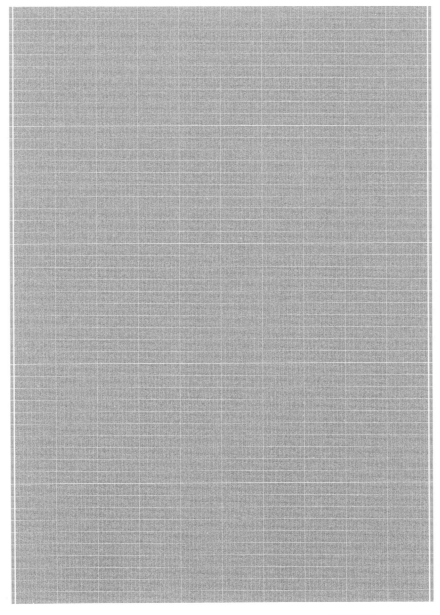

π upto 100,000,000 decimal digits　　　　　13000001–13500000

13500001–14000000 円周率 100,000,000 桁表

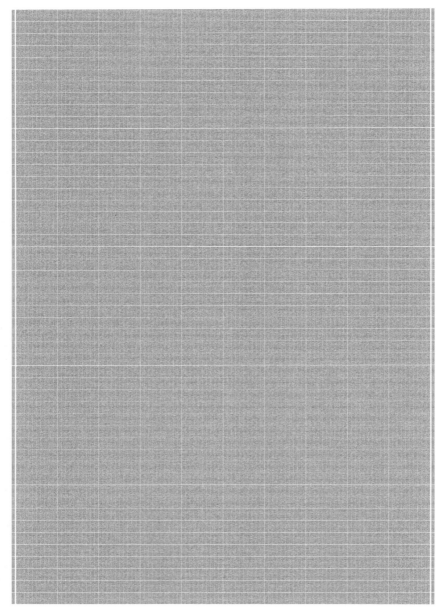

13500001–14000000 π upto 100,000,000 decimal digits

円周率 100,000,000 桁表　　　　　14000001-14500000

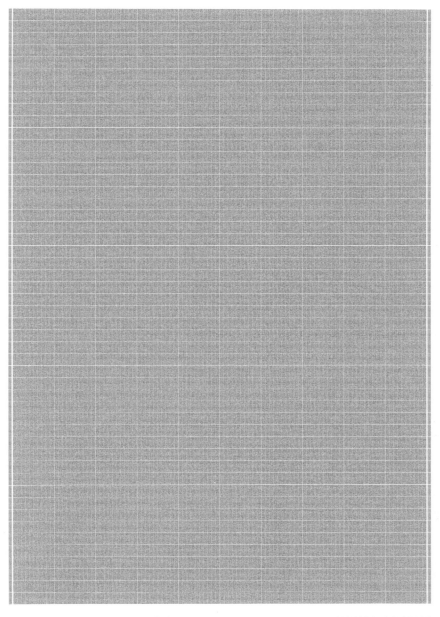

π upto 100,000,000 decimal digits　　　　14000001-14500000

14500001–15000000 円周率 100,000,000 桁表

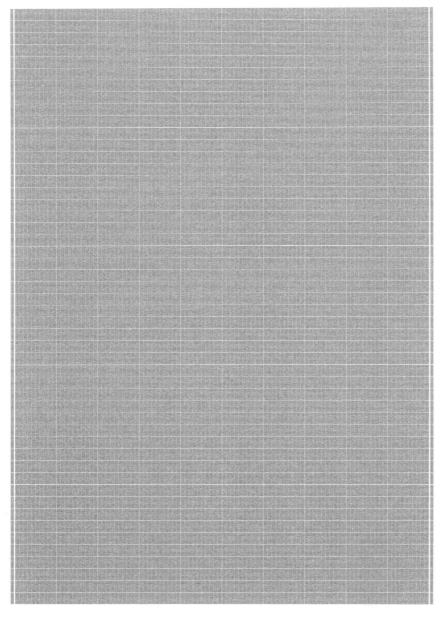

14500001–15000000 π upto 100,000,000 decimal digits

円周率 100,000,000 桁表　　　　　　　　　　　15000001–15500000

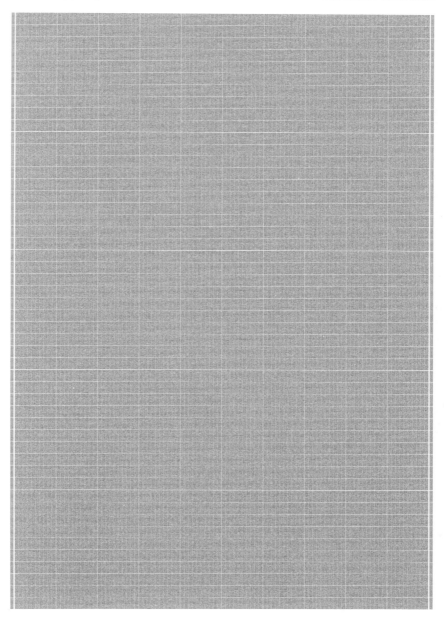

π upto 100,000,000 decimal digits　　　　　　15000001–15500000

15500001–16000000 円周率 100,000,000 桁表

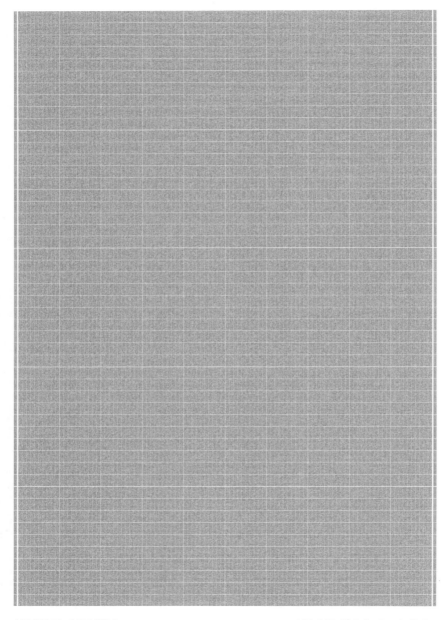

15500001–16000000 π upto 100,000,000 decimal digits

円周率 100,000,000 桁表　　　　　　　　　16000001–16500000

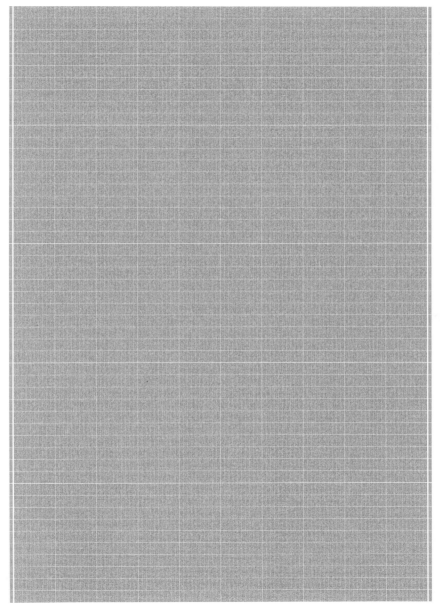

π upto 100,000,000 decimal digits　　　　　　16000001–16500000

16500001-17000000 円周率 100,000,000 桁表

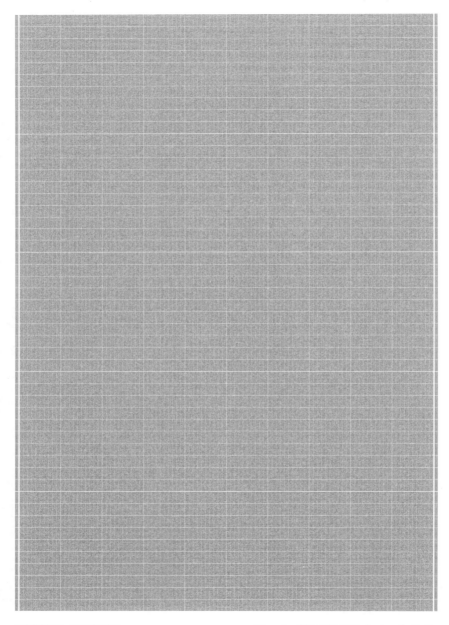

16500001-17000000 π upto 100,000,000 decimal digits

円周率 100,000,000 桁表 17000001–17500000

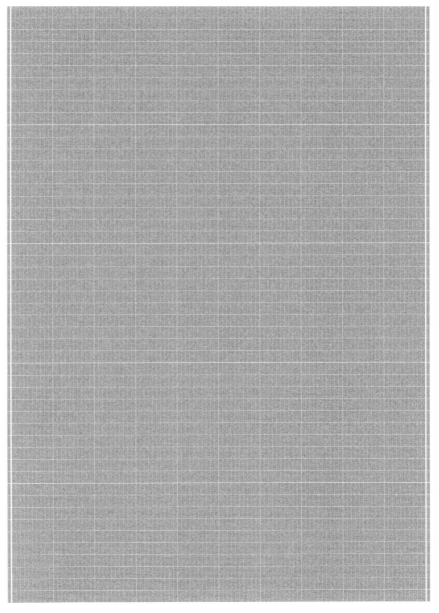

π upto 100,000,000 decimal digits 17000001–17500000

17500001–18000000 円周率 100,000,000 桁表

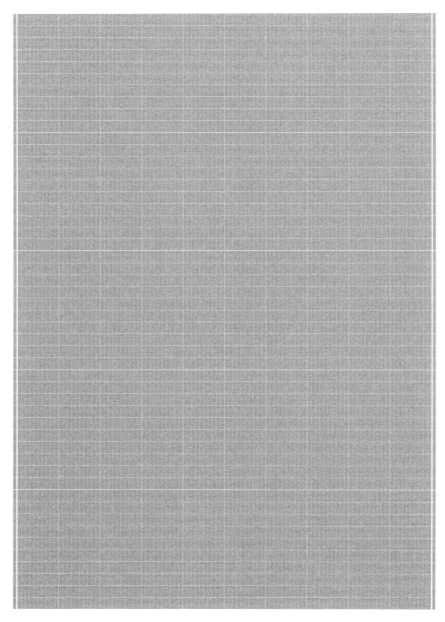

17500001–18000000 π upto 100,000,000 decimal digits

円周率 100,000,000 桁表　　　　　　　　　　　　18000001–18500000

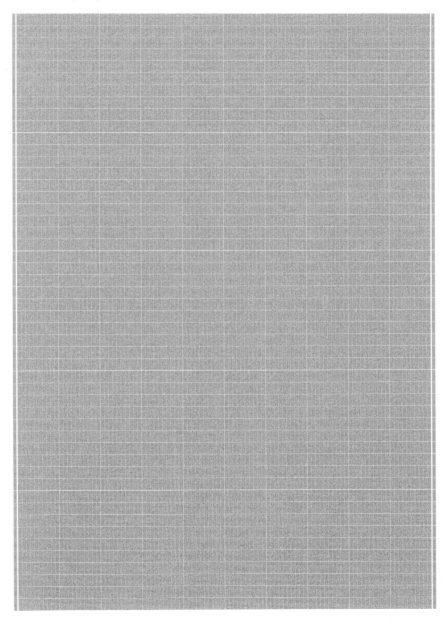

π upto 100,000,000 decimal digits　　　　　　　18000001–18500000

18500001–19000000 円周率 100,000,000 桁表

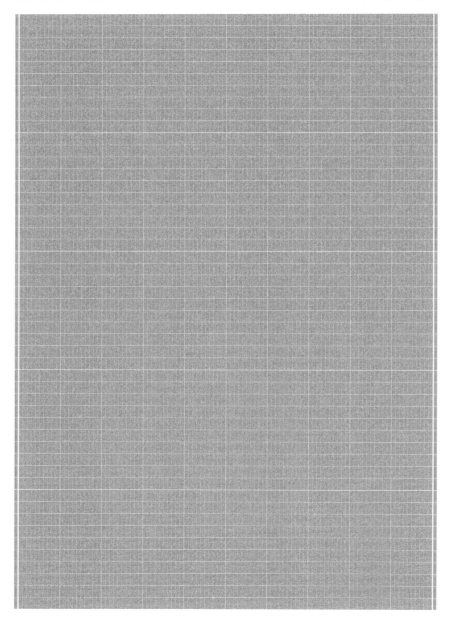

18500001–19000000 π upto 100,000,000 decimal digits

円周率 100,000,000 桁表　　　19000001–19500000

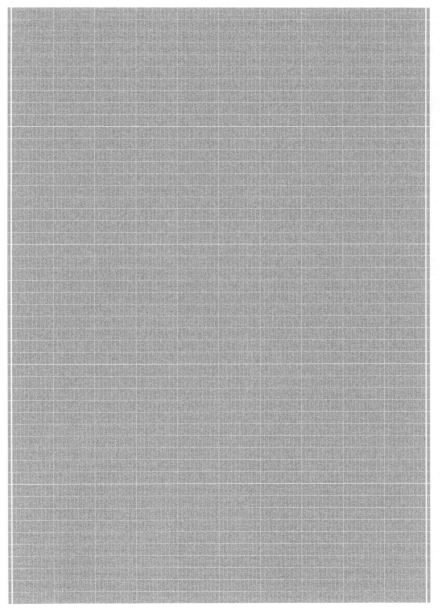

π upto 100,000,000 decimal digits　　　19000001–19500000

19500001–20000000 円周率 100,000,000 桁表

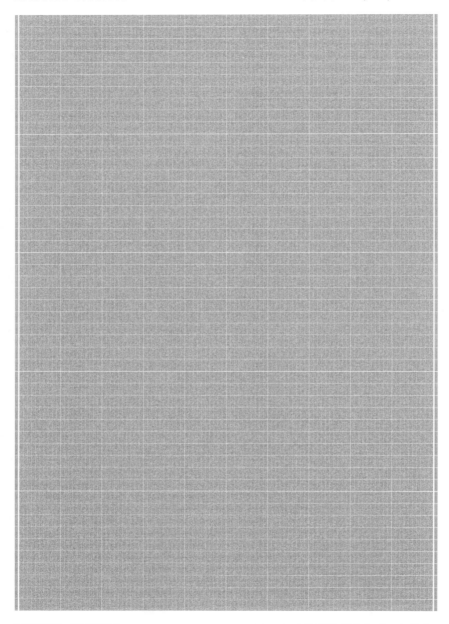

19500001–20000000 π upto 100,000,000 decimal digits

円周率 100,000,000 桁表 20000001–20500000

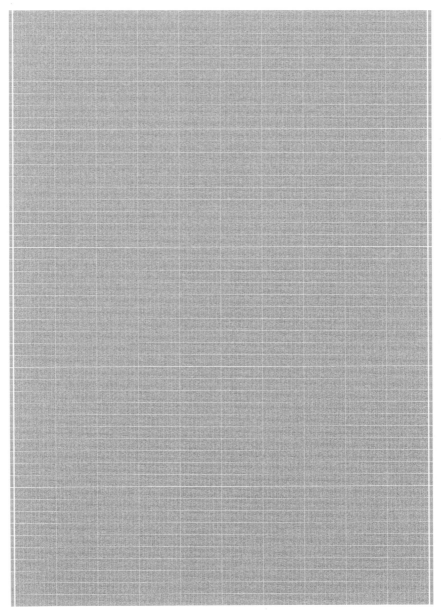

π upto 100,000,000 decimal digits 20000001–20500000

20500001–21000000 円周率 100,000,000 桁表

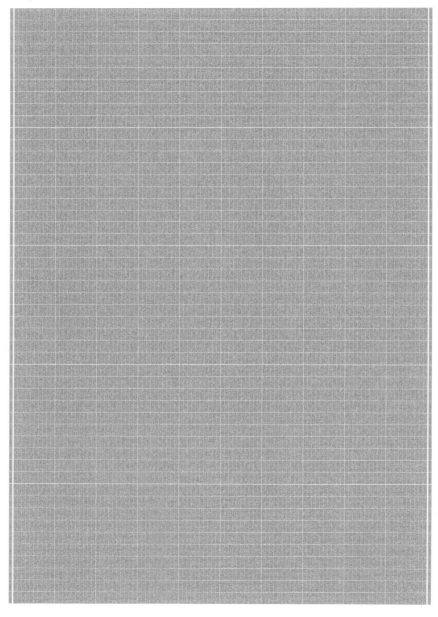

20500001–21000000 π upto 100,000,000 decimal digits

円周率 100,000,000 桁表　　　　　　　　21000001–21500000

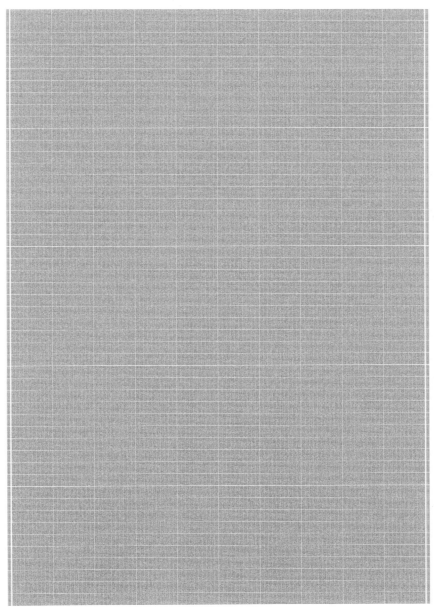

π upto 100,000,000 decimal digits　　　　　21000001–21500000

21500001–22000000 円周率 100,000,000 桁表

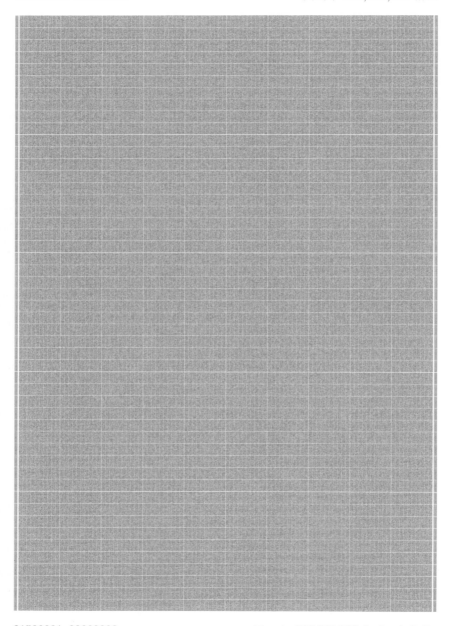

21500001–22000000 π upto 100,000,000 decimal digits

円周率 100,000,000 桁表　　　　　　　　　22000001–22500000

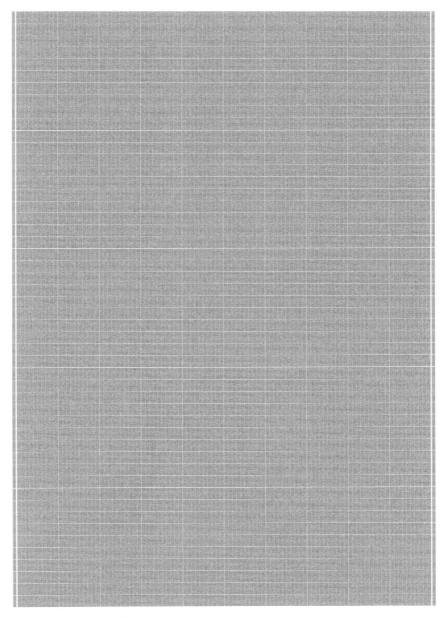

π upto 100,000,000 decimal digits　　　　　　22000001–22500000

22500001–23000000 円周率 100,000,000 桁表

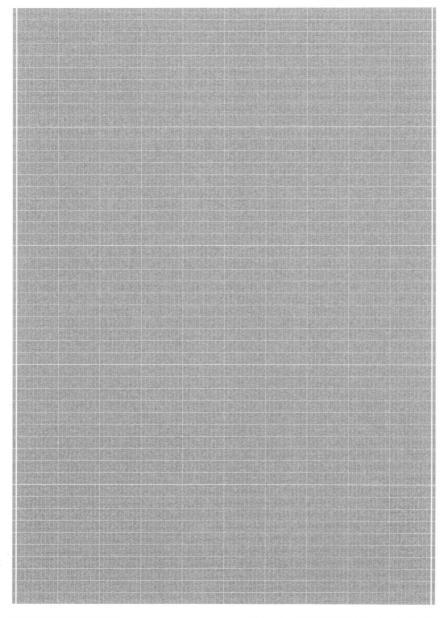

22500001–23000000 π upto 100,000,000 decimal digits

円周率 100,000,000 桁表　　23000001–23500000

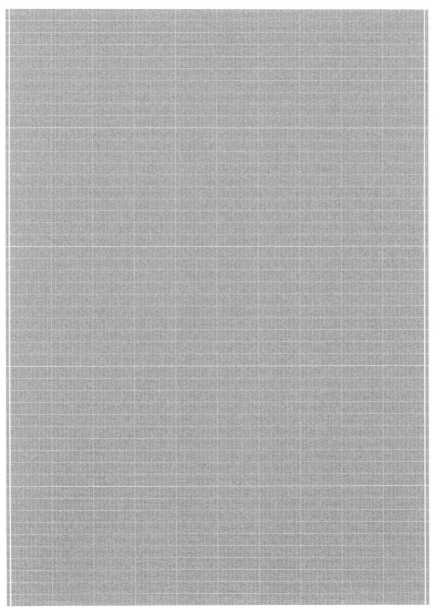

π upto 100,000,000 decimal digits　　23000001–23500000

23500001–24000000 円周率 100,000,000 桁表

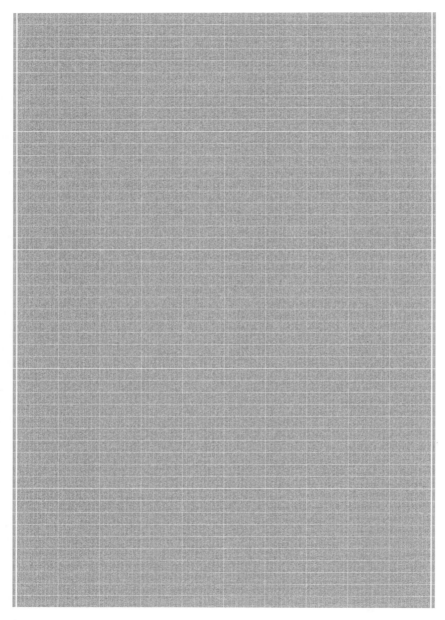

23500001–24000000 π upto 100,000,000 decimal digits

円周率 100,000,000 桁表 24000001–24500000

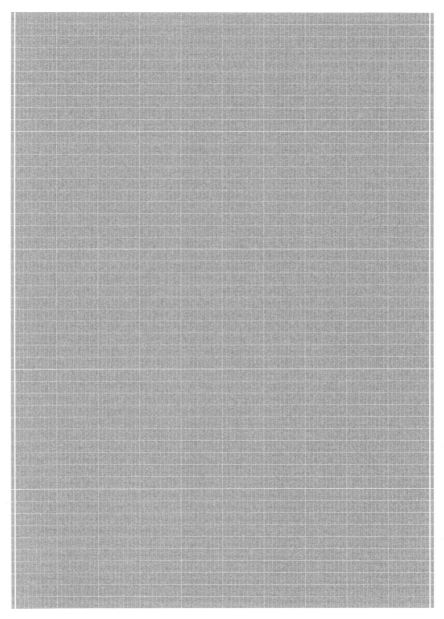

π upto 100,000,000 decimal digits 24000001–24500000

24500001–25000000 円周率 100,000,000 桁表

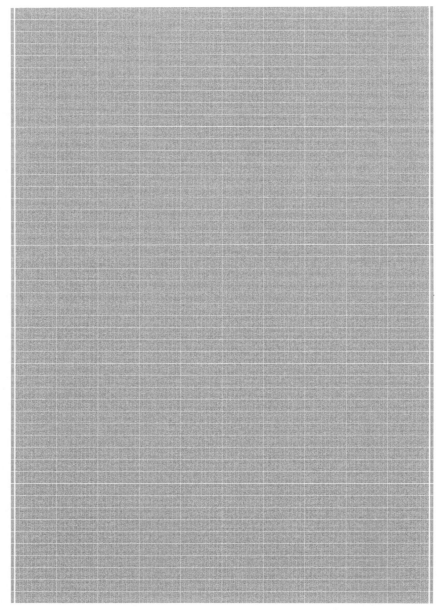

24500001–25000000 π upto 100,000,000 decimal digits

円周率 100,000,000 桁表 25000001–25500000

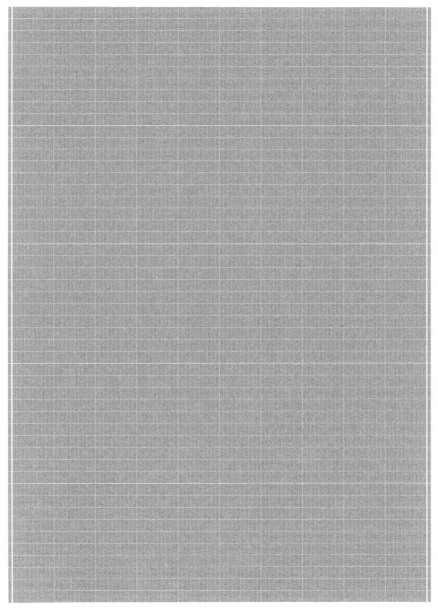

π upto 100,000,000 decimal digits 25000001–25500000

円周率 100,000,000 桁表　　　　　　　　26000001–26500000

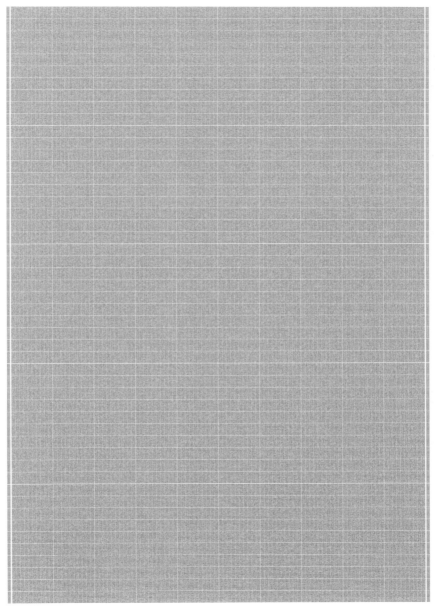

π upto 100,000,000 decimal digits　　　　26000001–26500000

26500001–27000000 円周率 100,000,000 桁表

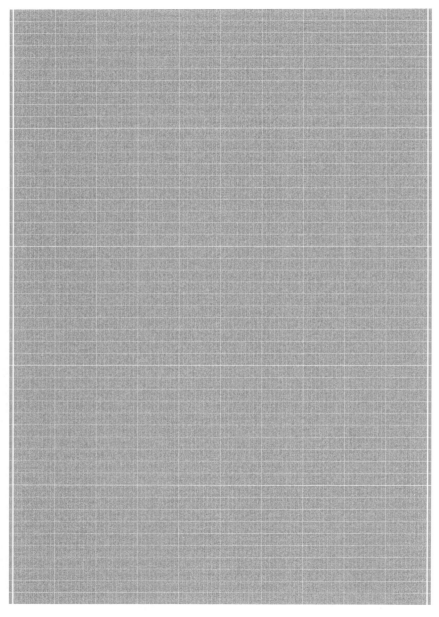

26500001–27000000 π upto 100,000,000 decimal digits

円周率 100,000,000 桁表　　　　　27000001–27500000

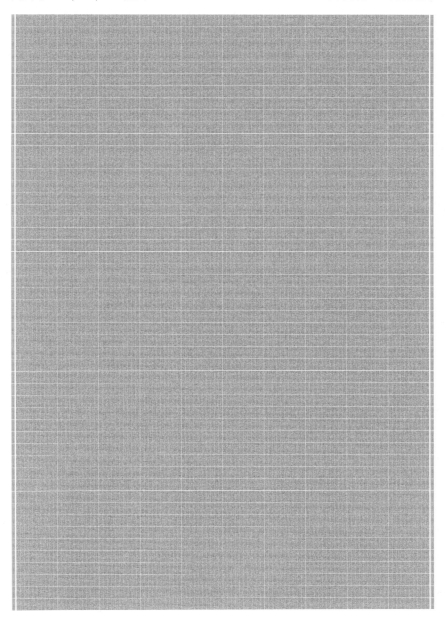

π upto 100,000,000 decimal digits　　　　27000001–27500000

27500001–28000000 円周率 100,000,000 桁表

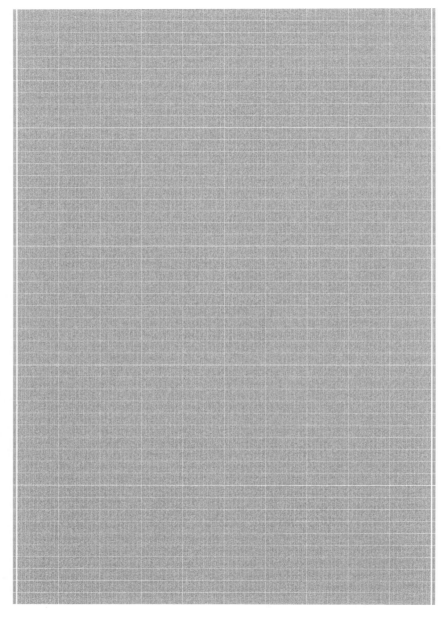

27500001–28000000 π upto 100,000,000 decimal digits

円周率 100,000,000 桁表 28000001–28500000

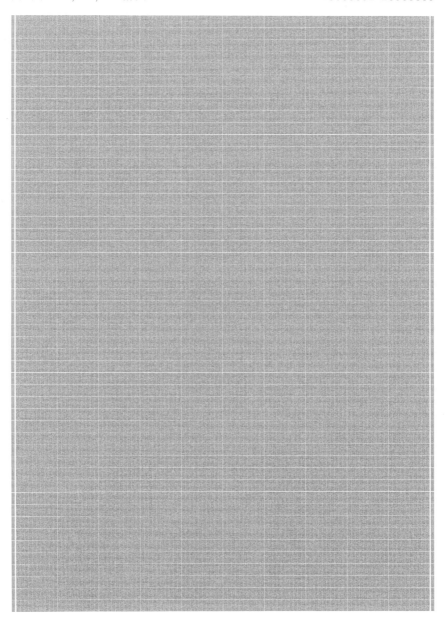

π upto 100,000,000 decimal digits 28000001–28500000

28500001–29000000 円周率 100,000,000 桁表

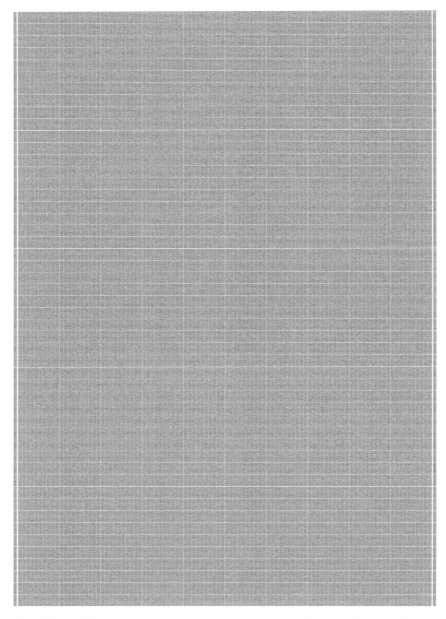

π upto 100,000,000 decimal digits

円周率 100,000,000 桁表 29000001–29500000

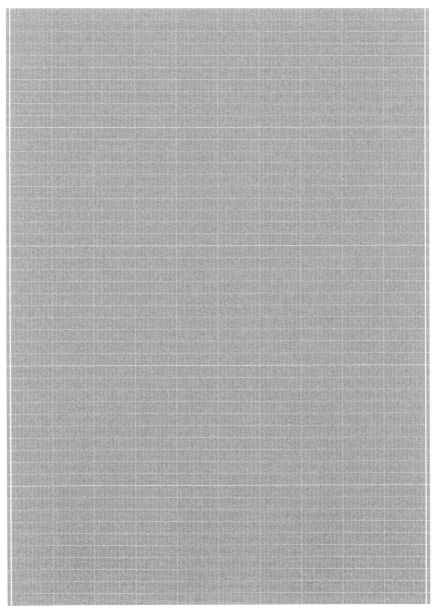

π upto 100,000,000 decimal digits 29000001–29500000

29500001–30000000 円周率 100,000,000 桁表

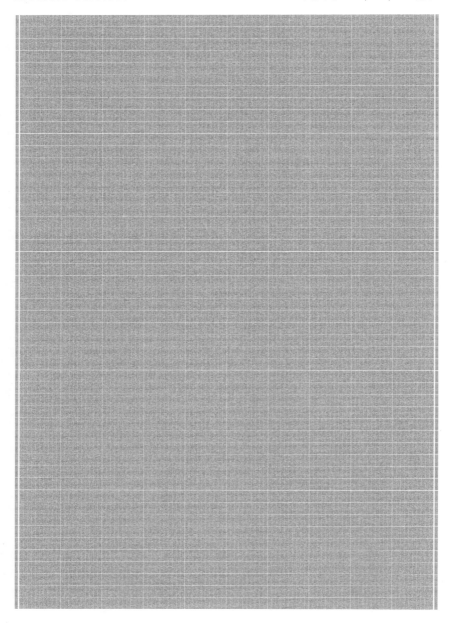

29500001–30000000 π upto 100,000,000 decimal digits

円周率 100,000,000 桁表 30000001–30500000

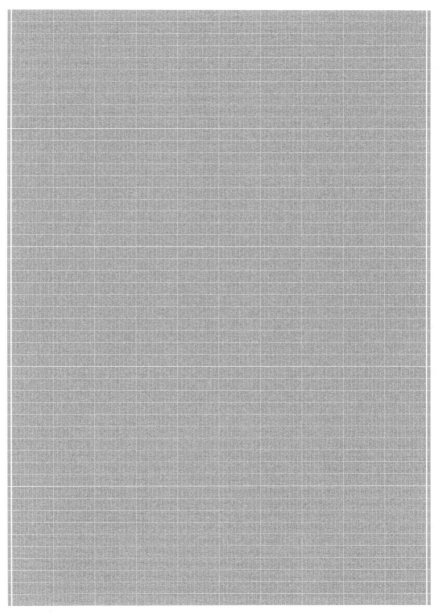

π upto 100,000,000 decimal digits 30000001–30500000

30500001–31000000 円周率 100,000,000 桁表

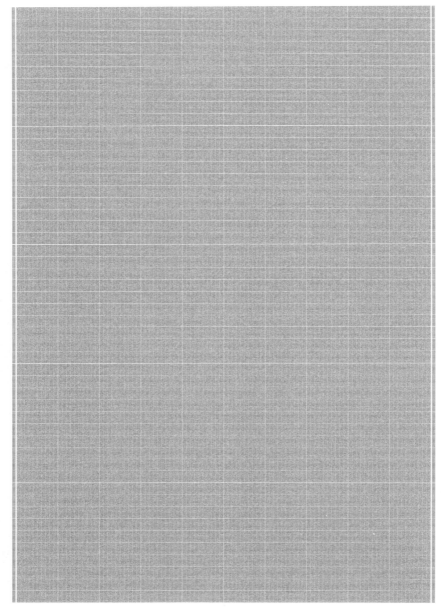

30500001–31000000 π upto 100,000,000 decimal digits

円周率 100,000,000 桁表　　　31000001–31500000

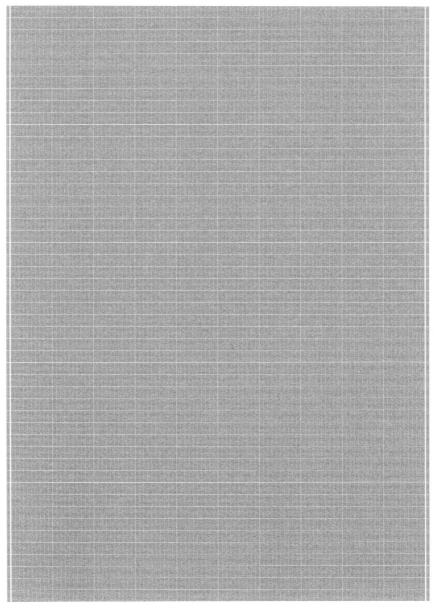

π upto 100,000,000 decimal digits　　　31000001–31500000

31500001–32000000 円周率 100,000,000 桁表

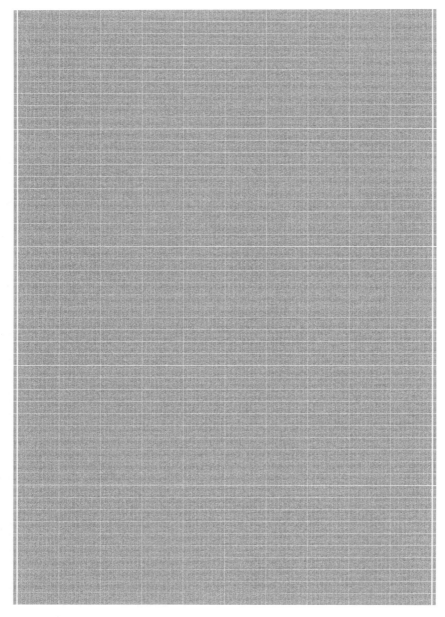

31500001–32000000 π upto 100,000,000 decimal digits

円周率 100,000,000 桁表 32000001–32500000

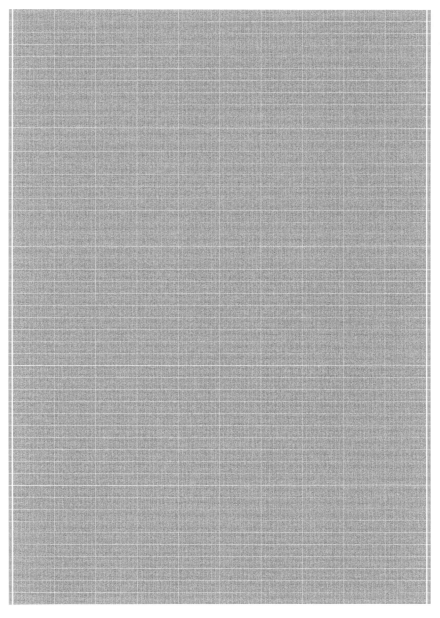

π upto 100,000,000 decimal digits 32000001–32500000

32500001–33000000 円周率 100,000,000 桁表

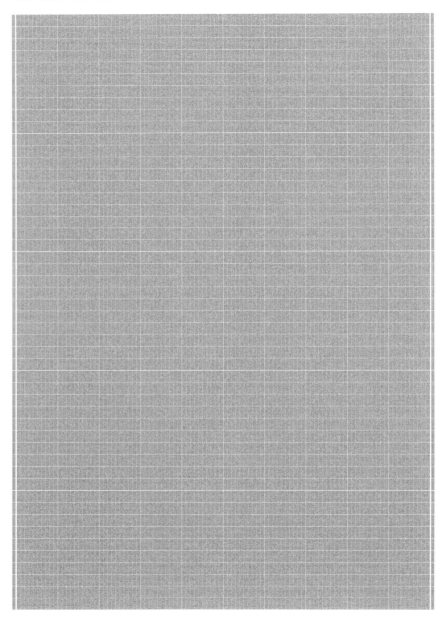

32500001–33000000 π upto 100,000,000 decimal digits

円周率 100,000,000 桁表 33000001–33500000

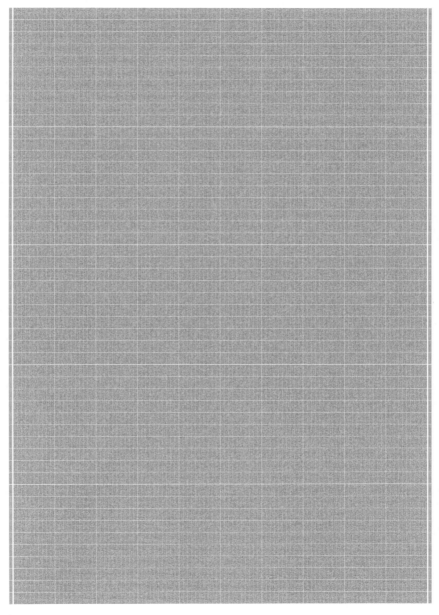

π upto 100,000,000 decimal digits 33000001–33500000

33500001–34000000 円周率 100,000,000 桁表

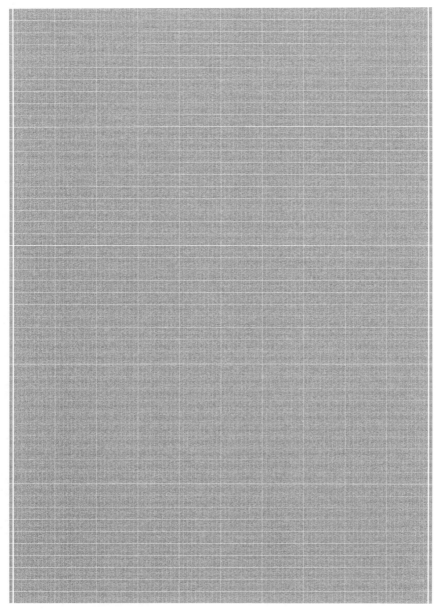

33500001–34000000 π upto 100,000,000 decimal digits

円周率 100,000,000 桁表　　　　　　　　　　34000001–34500000

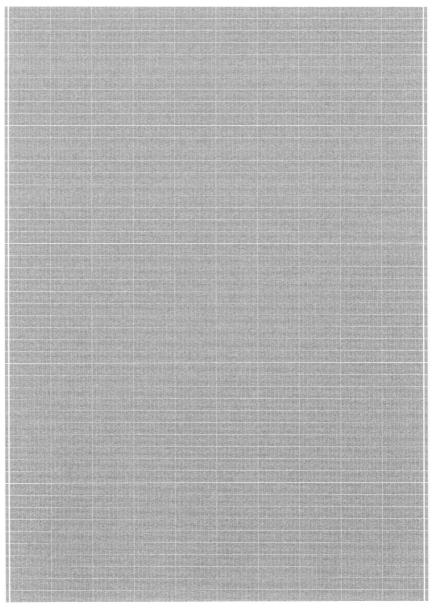

π upto 100,000,000 decimal digits　　　　34000001–34500000

34500001-35000000 円周率100,000,000桁表

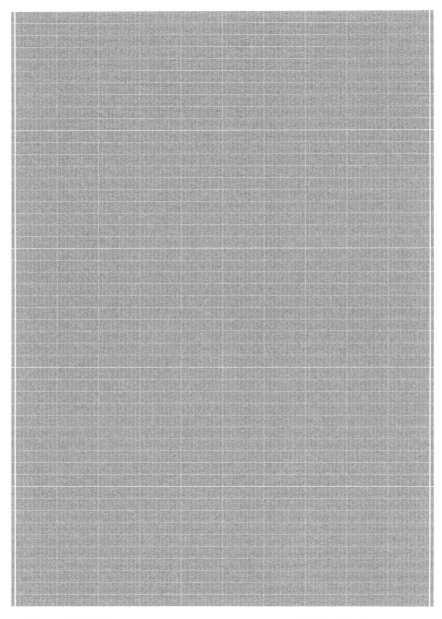

34500001-35000000 π upto 100,000,000 decimal digits

円周率 100,000,000 桁表 35000001–35500000

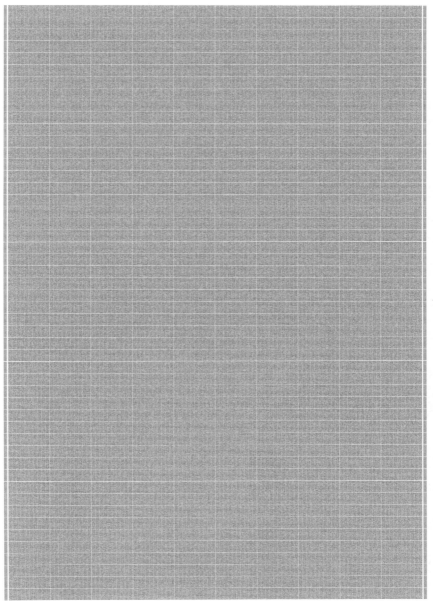

π upto 100,000,000 decimal digits 35000001–35500000

35500001-36000000 円周率100,000,000桁表

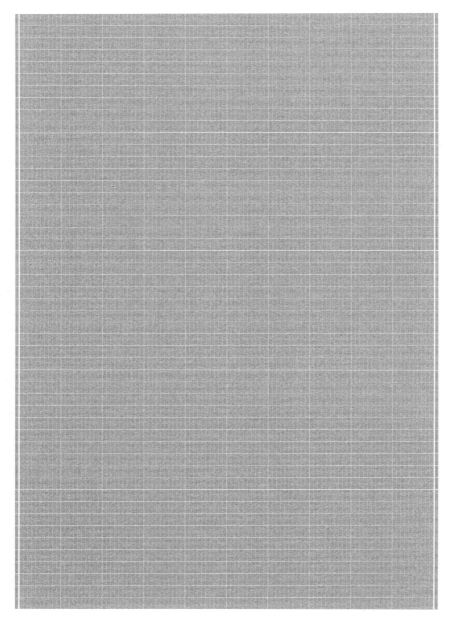

35500001-36000000 π upto 100,000,000 decimal digits

円周率 100,000,000 桁表 36000001–36500000

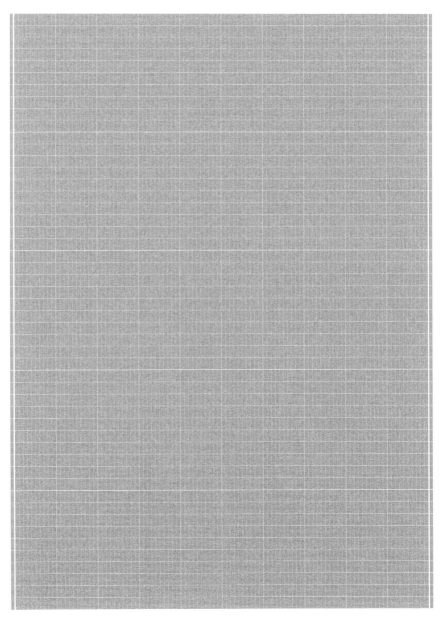

π upto 100,000,000 decimal digits 36000001–36500000

36500001-37000000 円周率 100,000,000 桁表

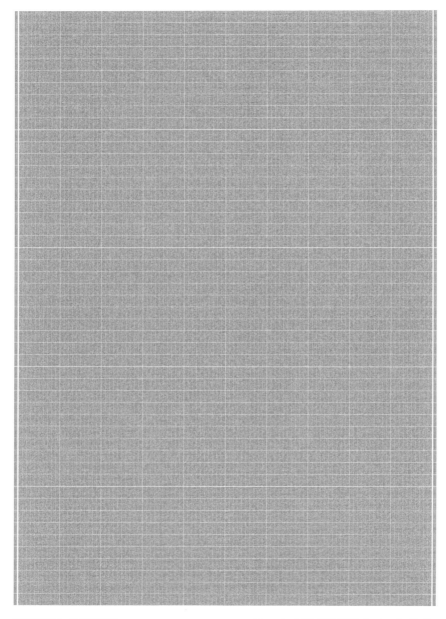

36500001-37000000 π upto 100,000,000 decimal digits

円周率 100,000,000 桁表 37000001–37500000

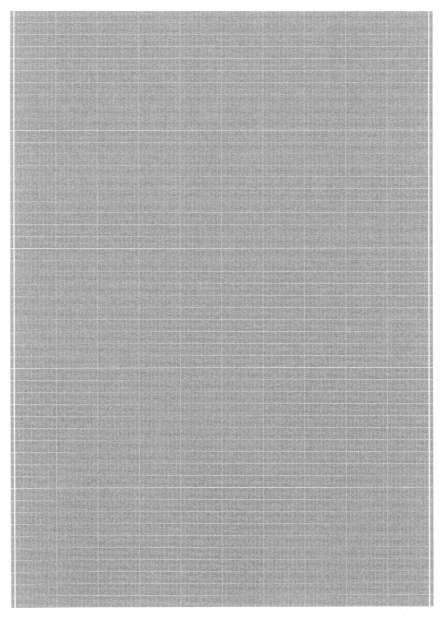

π upto 100,000,000 decimal digits 37000001–37500000

37500001–38000000 円周率100,000,000桁表

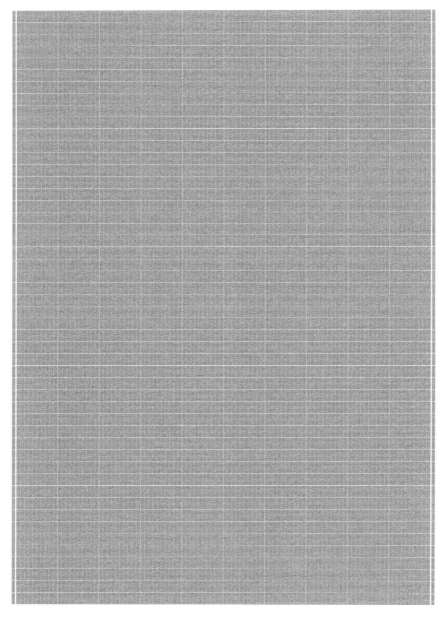

37500001–38000000 π upto 100,000,000 decimal digits

円周率 100,000,000 桁表 38000001–38500000

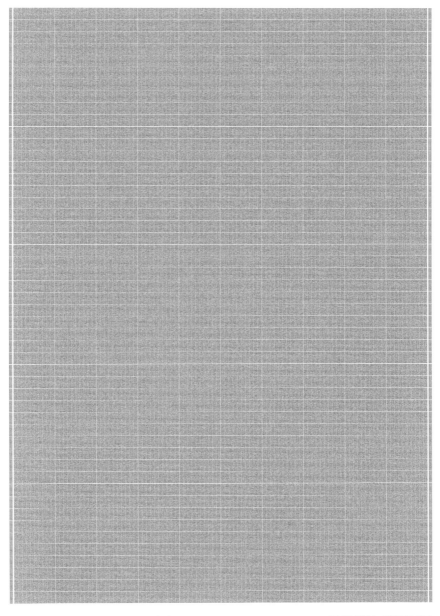

π upto 100,000,000 decimal digits 38000001–38500000

38500001–39000000 円周率 100,000,000 桁表

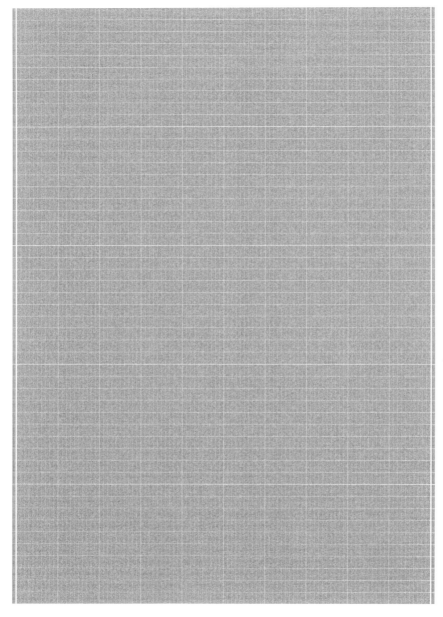

π upto 100,000,000 decimal digits

円周率 100,000,000 桁表　　　　　　　　39000001–39500000

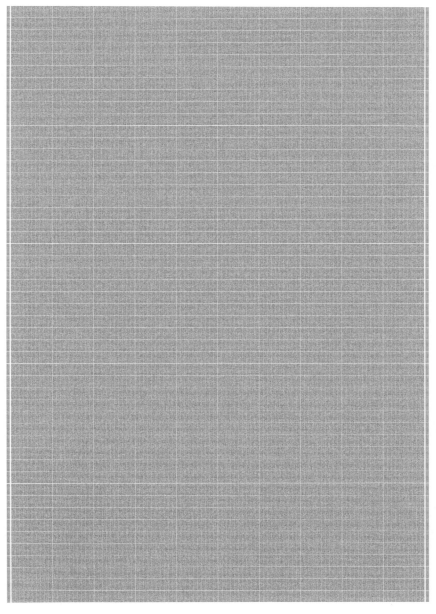

π upto 100,000,000 decimal digits　　　　　　39000001–39500000

39500001–40000000 円周率 100,000,000 桁表

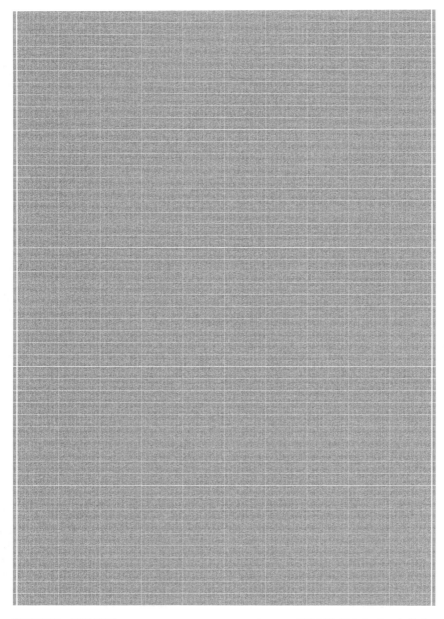

39500001–40000000 π upto 100,000,000 decimal digits

円周率 100,000,000 桁表　　　　　　　40000001–40500000

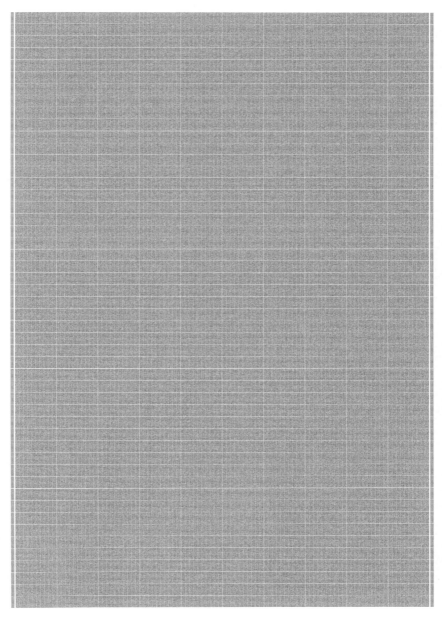

π upto 100,000,000 decimal digits　　　　　40000001–40500000

40500001–41000000 円周率 100,000,000 桁表

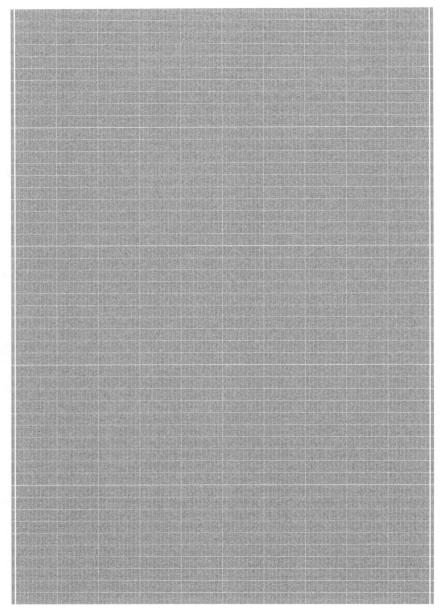

40500001–41000000 π upto 100,000,000 decimal digits

円周率 100,000,000 桁表　　　41000001–41500000

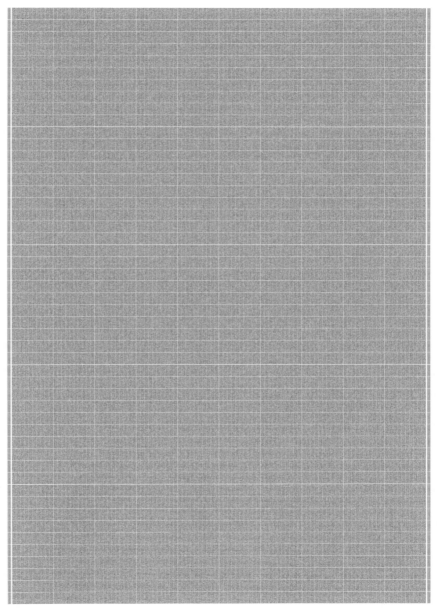

π upto 100,000,000 decimal digits　　　41000001–41500000

41500001–42000000 円周率 100,000,000 桁表

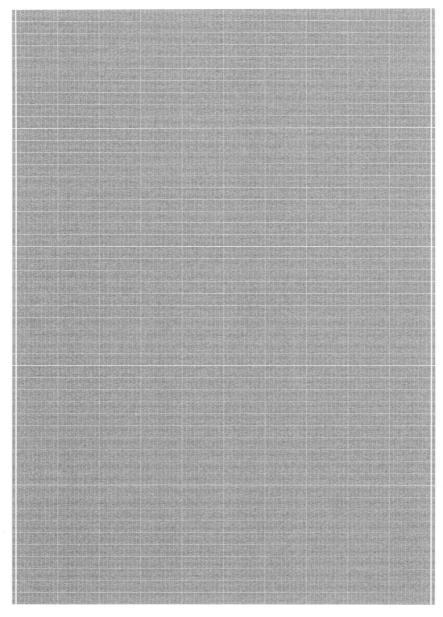

41500001–42000000 π upto 100,000,000 decimal digits

円周率 100,000,000 桁表 42000001–42500000

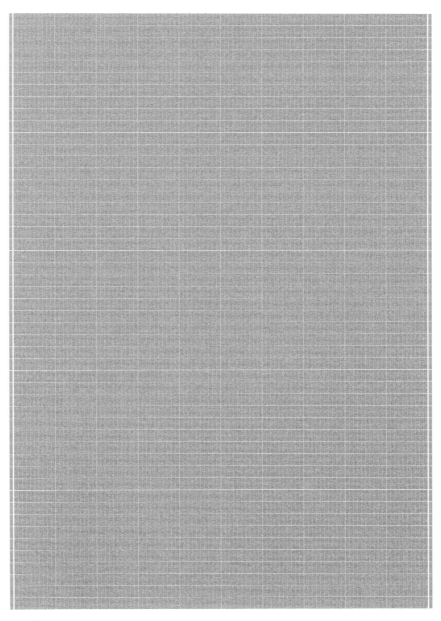

π upto 100,000,000 decimal digits 42000001–42500000

42500001–43000000 円周率 100,000,000 桁表

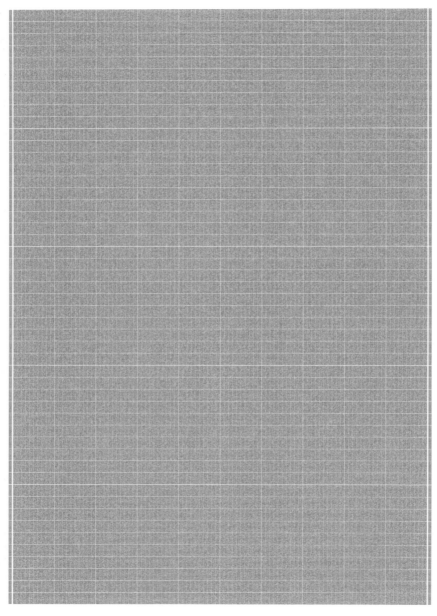

42500001–43000000 π upto 100,000,000 decimal digits

円周率 100,000,000 桁表　　　　　　　　　43000001–43500000

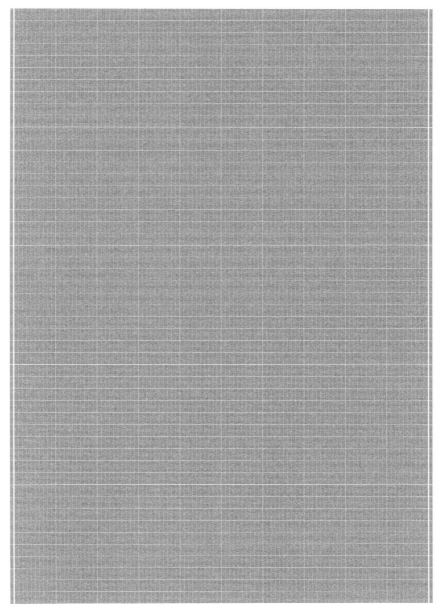

π upto 100,000,000 decimal digits　　　　　43000001–43500000

43500001–44000000 円周率 100,000,000 桁表

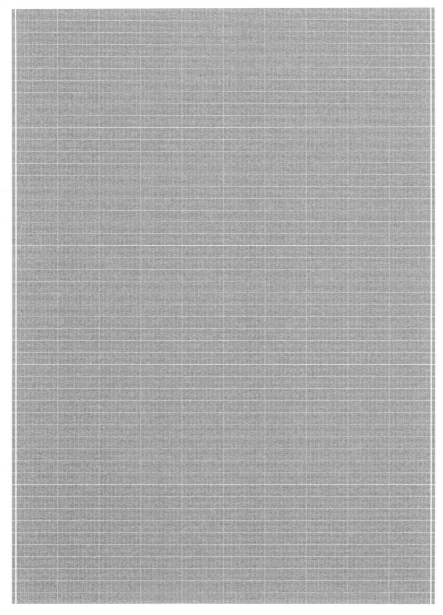

43500001–44000000 π upto 100,000,000 decimal digits

円周率 100,000,000 桁表 44000001–44500000

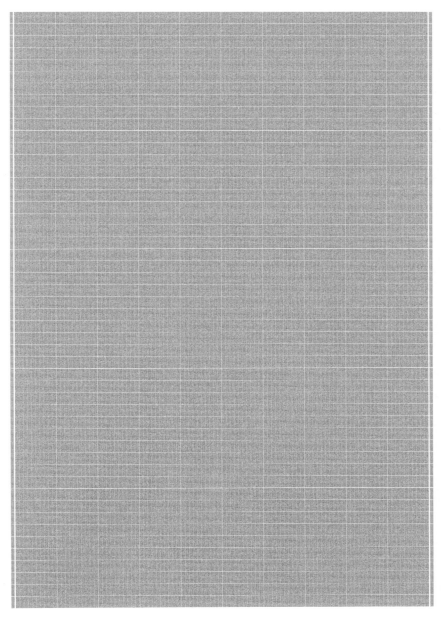

π upto 100,000,000 decimal digits 44000001–44500000

円周率 100,000,000 桁表　　　　　　　　　　　45000001–45500000

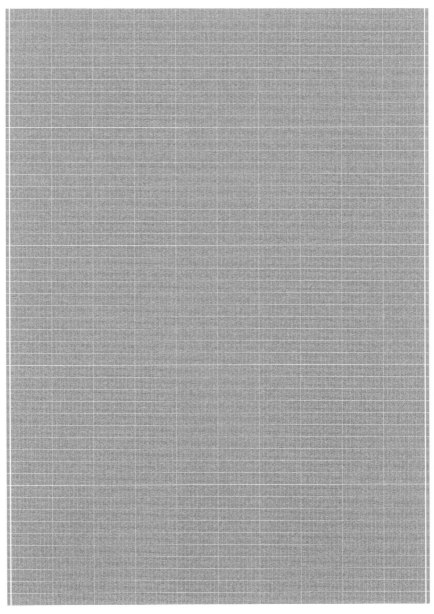

π upto 100,000,000 decimal digits　　　　　　45000001–45500000

45500001–46000000 円周率 100,000,000 桁表

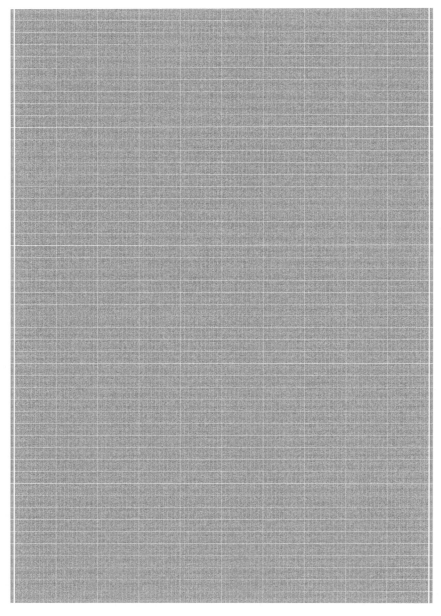

45500001–46000000 π upto 100,000,000 decimal digits

円周率 100,000,000 桁表　　　　　　　　　　　　46000001–46500000

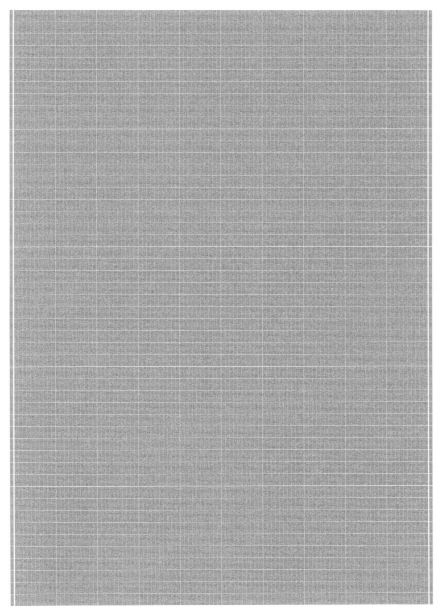

π upto 100,000,000 decimal digits　　　　　　　46000001–46500000

46500001–47000000 円周率 100,000,000 桁表

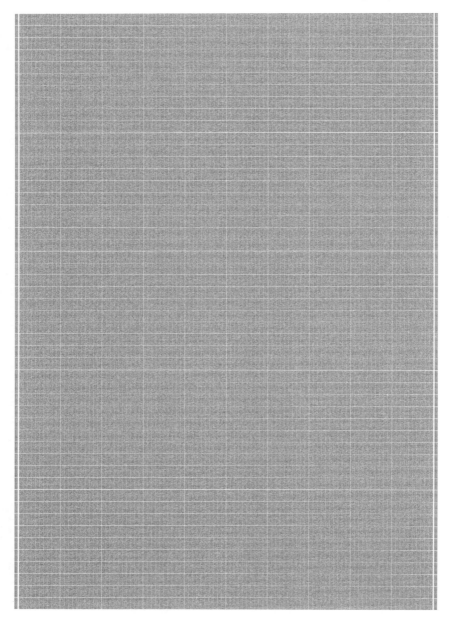

46500001–47000000 π upto 100,000,000 decimal digits

円周率 100,000,000 桁表　　　　　　　　　47000001–47500000

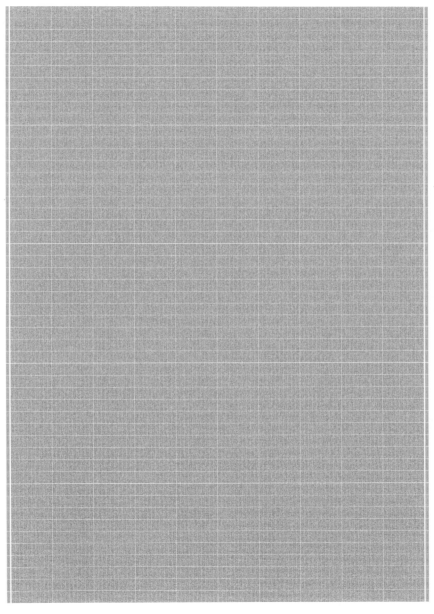

π upto 100,000,000 decimal digits　　　　47000001–47500000

47500001–48000000 円周率 100,000,000 桁表

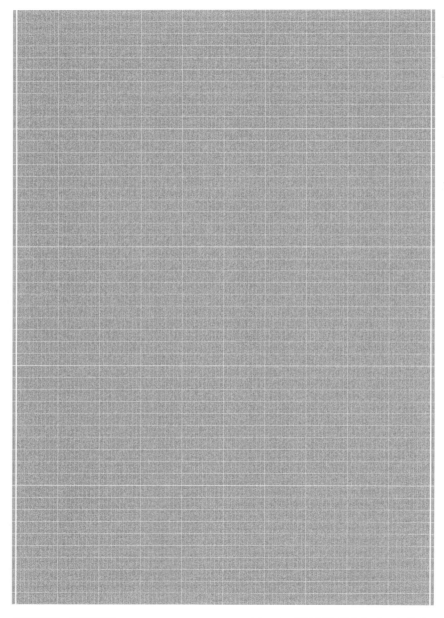

47500001–48000000 π upto 100,000,000 decimal digits

円周率 100,000,000 桁表　　　　　　　　　48000001–48500000

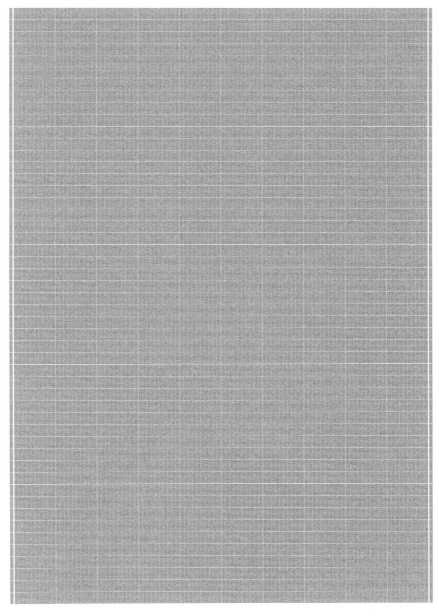

π upto 100,000,000 decimal digits　　　　48000001–48500000

48500001–49000000 円周率 100,000,000 桁表

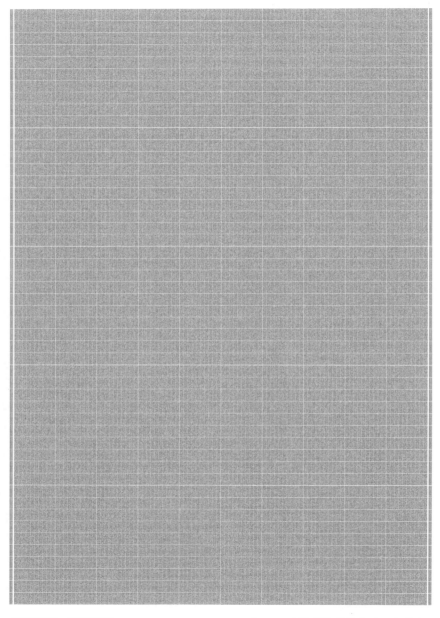

48500001–49000000 π upto 100,000,000 decimal digits

円周率 100,000,000 桁表　　　49000001–49500000

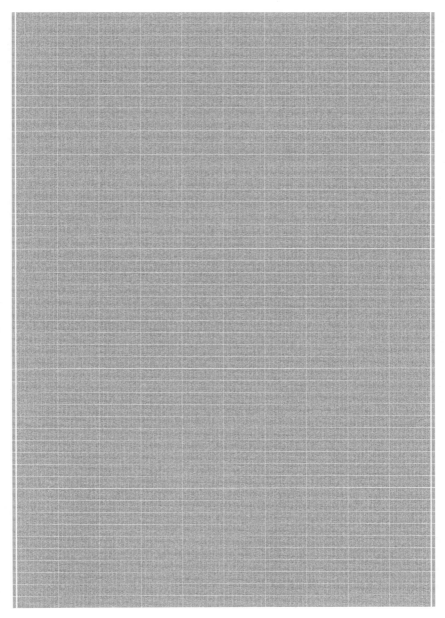

π upto 100,000,000 decimal digits　　　49000001–49500000

49500001-50000000 円周率 100,000,000 桁表

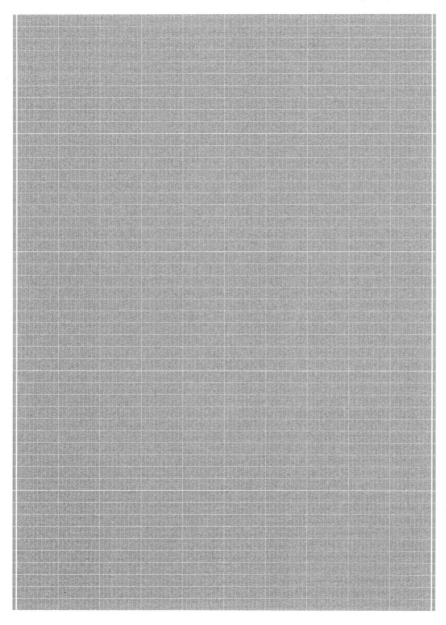

49500001-50000000 π upto 100,000,000 decimal digits

円周率 100,000,000 桁表　　　　　　　　　　　50000001–50500000

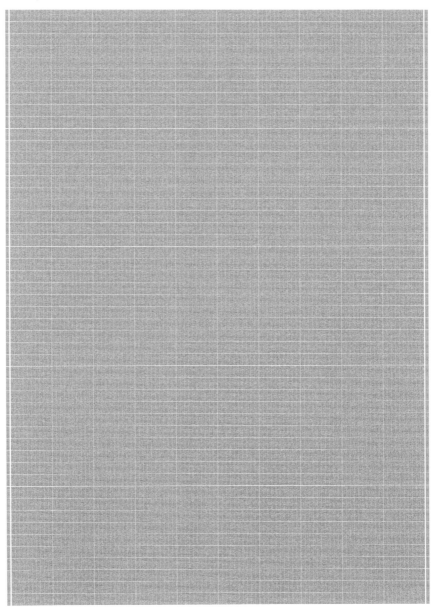

π upto 100,000,000 decimal digits　　　　　　50000001–50500000

50500001-51000000 円周率 100,000,000 桁表

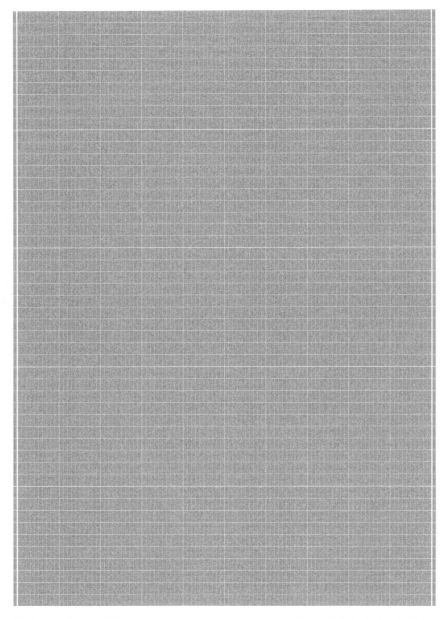

50500001-51000000　　　　　π upto 100,000,000 decimal digits

円周率 100,000,000 桁表　　　　　　　　51000001–51500000

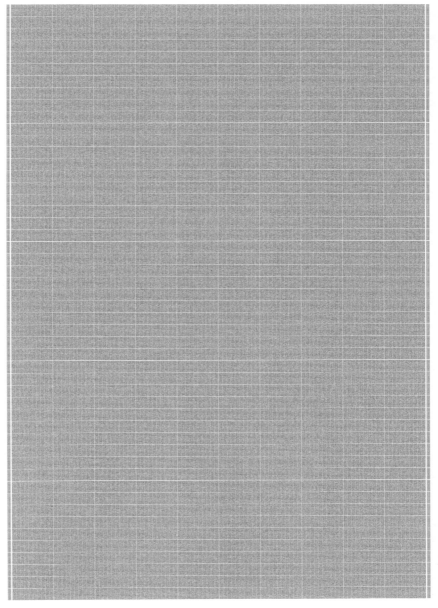

π upto 100,000,000 decimal digits　　　　　　51000001–51500000

51500001–52000000 円周率 100,000,000 桁表

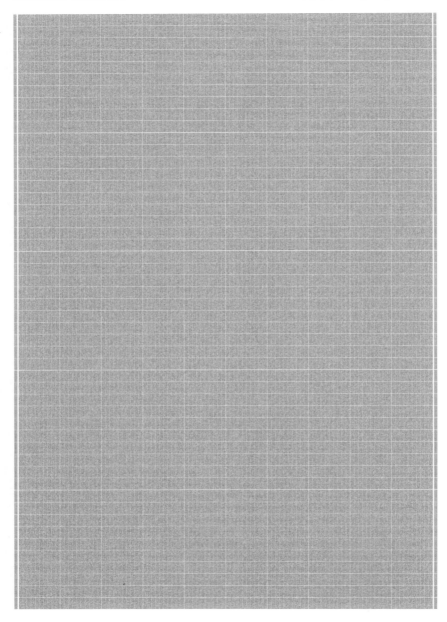

51500001–52000000 π upto 100,000,000 decimal digits

円周率 100,000,000 桁表　　　　　　52000001–52500000

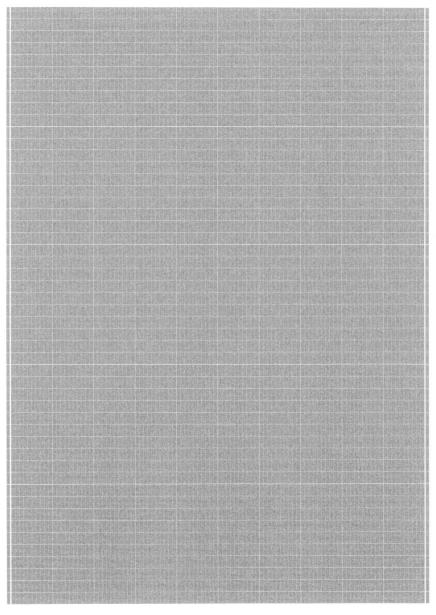

π upto 100,000,000 decimal digits　　　52000001–52500000

52500001–53000000 円周率 100,000,000 桁表

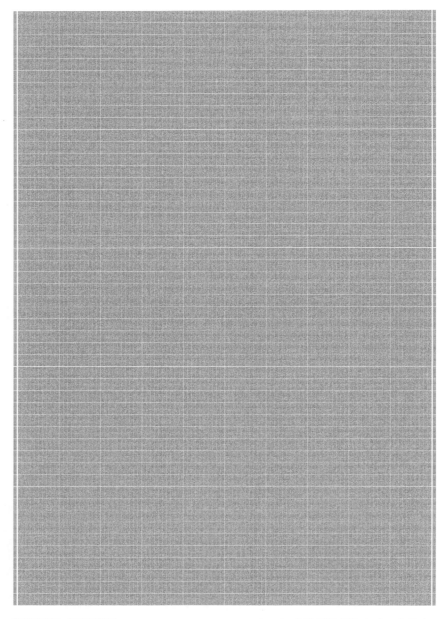

52500001–53000000 π upto 100,000,000 decimal digits

円周率 100,000,000 桁表　　　　　53000001–53500000

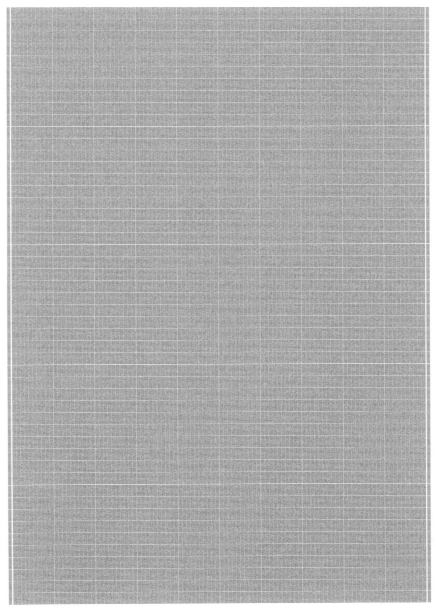

π upto 100,000,000 decimal digits　　　　　53000001–53500000

53500001–54000000 円周率 100,000,000 桁表

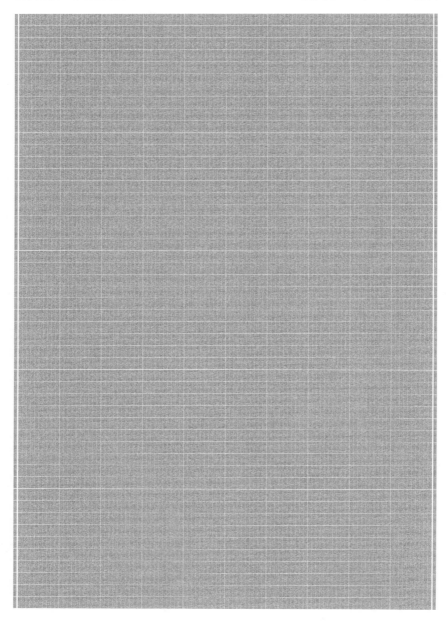

53500001–54000000 π upto 100,000,000 decimal digits

円周率 100,000,000 桁表 54000001–54500000

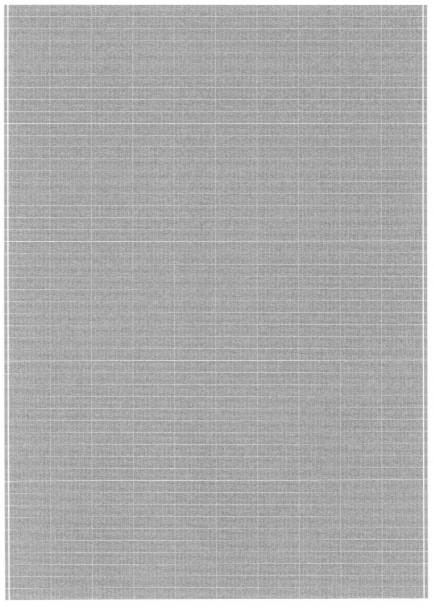

π upto 100,000,000 decimal digits 54000001–54500000

54500001–55000000 円周率100,000,000桁表

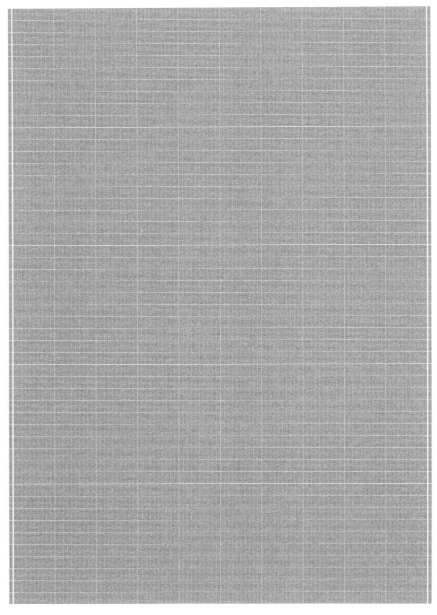

54500001–55000000 π upto 100,000,000 decimal digits

円周率 100,000,000 桁表 55000001–55500000

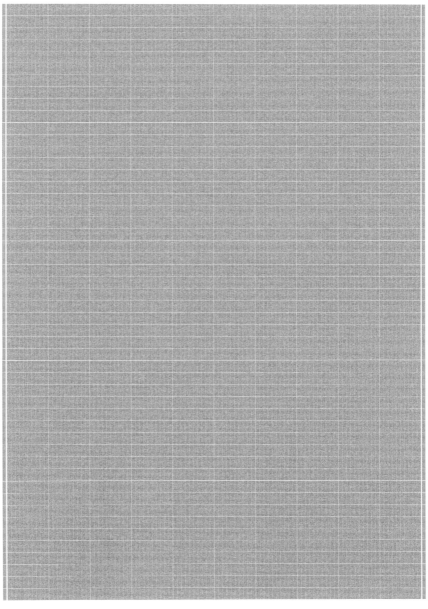

π upto 100,000,000 decimal digits 55000001–55500000

55500001–56000000 円周率 100,000,000 桁表

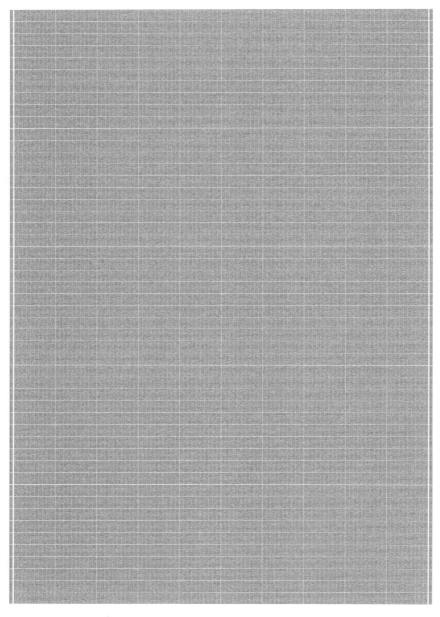

55500001–56000000 π upto 100,000,000 decimal digits

円周率 100,000,000 桁表　　　　　　　　56000001–56500000

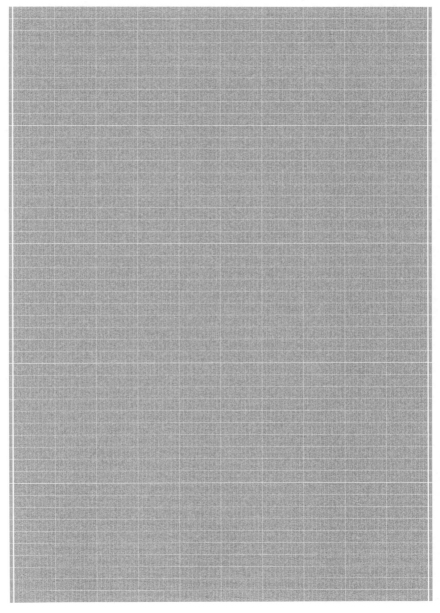

π upto 100,000,000 decimal digits　　　　　　　56000001–56500000

56500001–57000000　　　　　　　　　　　円周率 100,000,000 桁表

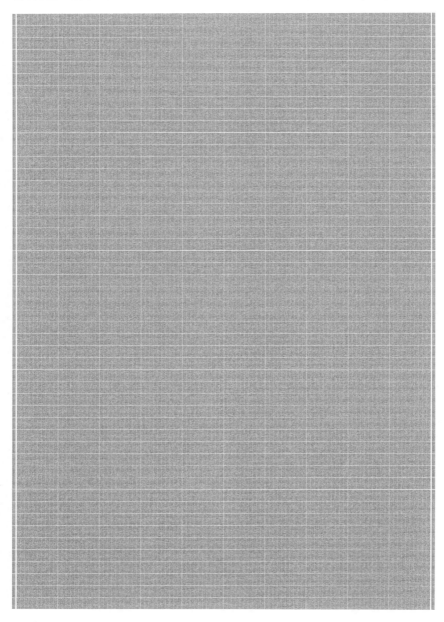

56500001–57000000　　　　　　　　　　π upto 100,000,000 decimal digits

円周率 100,000,000 桁表　　　　　　　　57000001–57500000

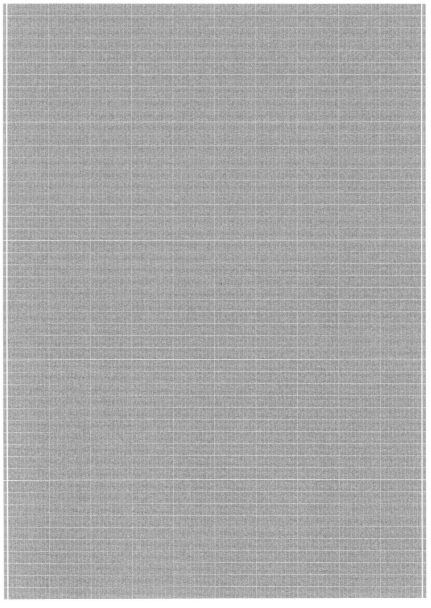

π upto 100,000,000 decimal digits　　　　57000001–57500000

57500001–58000000 円周率 100,000,000 桁表

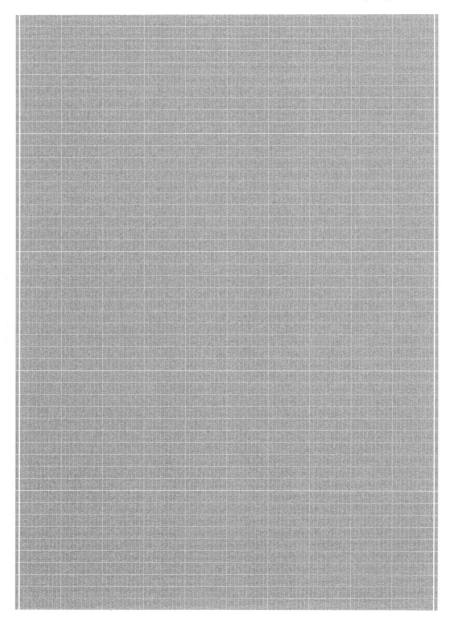

57500001–58000000 π upto 100,000,000 decimal digits

円周率 100,000,000 桁表　　　　　　　　　　58000001–58500000

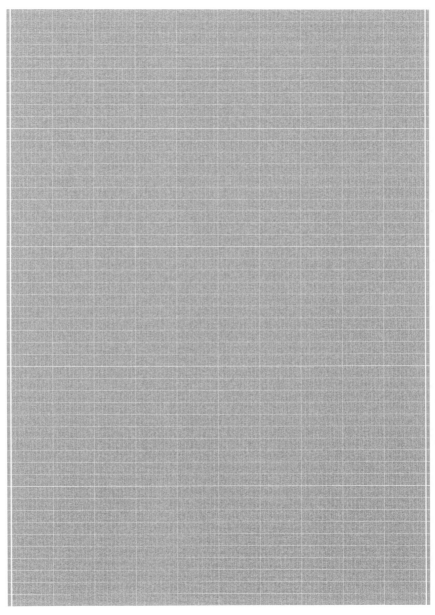

π upto 100,000,000 decimal digits　　　　　　58000001–58500000

58500001–59000000 円周率 100,000,000 桁表

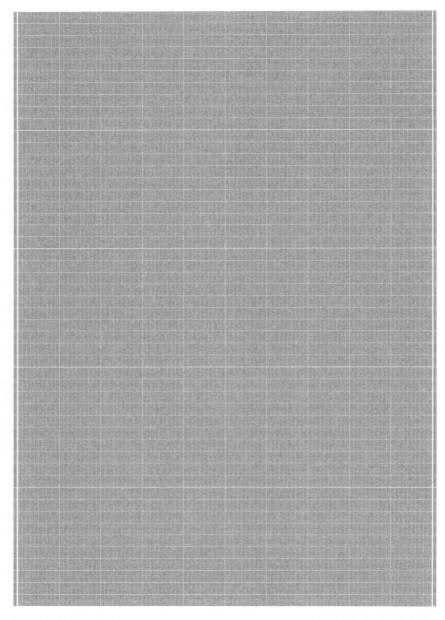

58500001–59000000 π upto 100,000,000 decimal digits

円周率 100,000,000 桁表 59000001–59500000

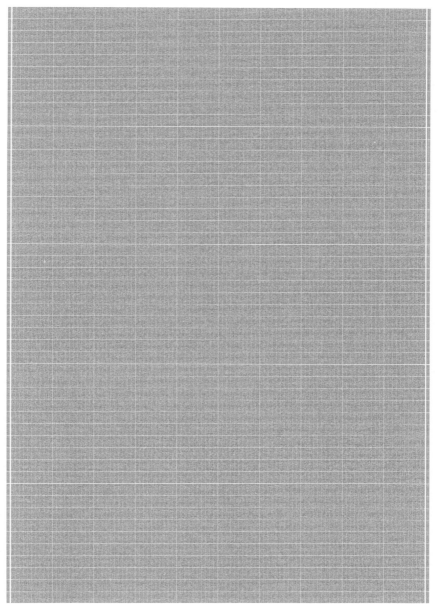

π upto 100,000,000 decimal digits 59000001–59500000

59500001–60000000 円周率 100,000,000 桁表

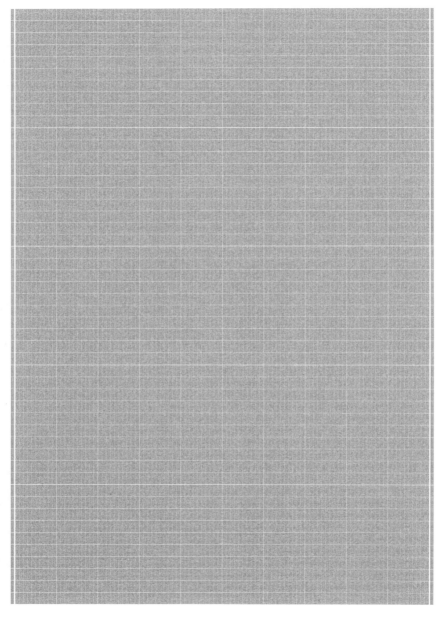

59500001–60000000 π upto 100,000,000 decimal digits

円周率 100,000,000 桁表　　　　　　　　　　　　60000001–60500000

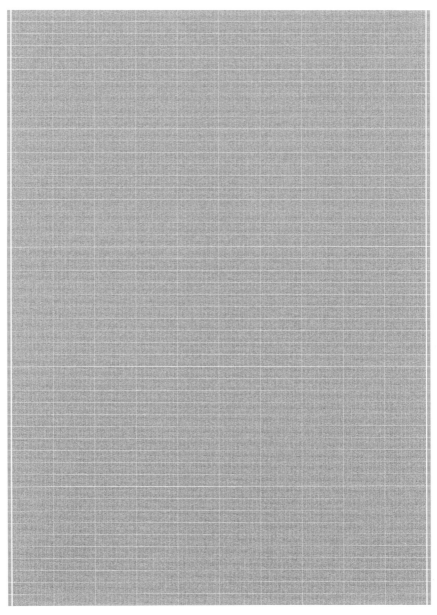

π upto 100,000,000 decimal digits　　　　　　60000001–60500000

60500001–61000000 円周率 100,000,000 桁表

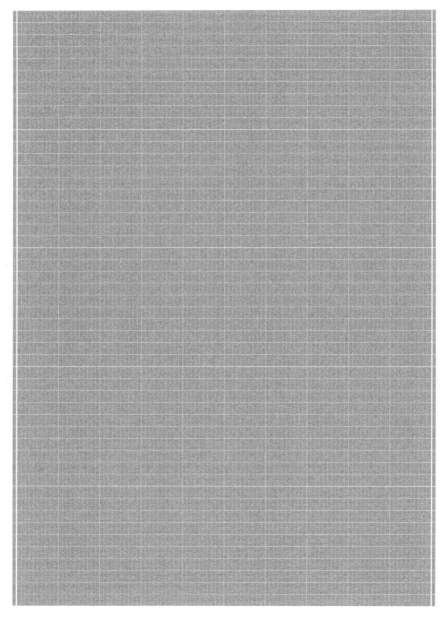

60500001–61000000 π upto 100,000,000 decimal digits

円周率 100,000,000 桁表　　　　　　　　61000001–61500000

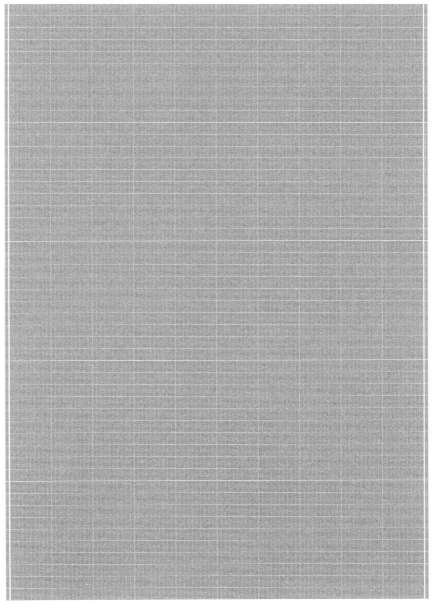

π upto 100,000,000 decimal digits　　　　61000001–61500000

61500001–62000000 円周率 100,000,000 桁表

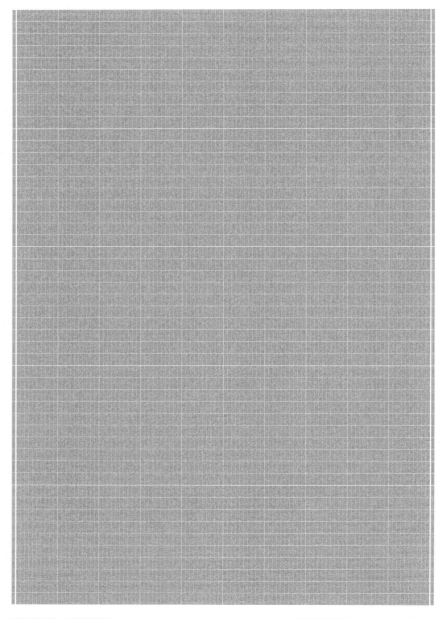

π upto 100,000,000 decimal digits

円周率 100,000,000 桁表　　　　　　　　　　62000001–62500000

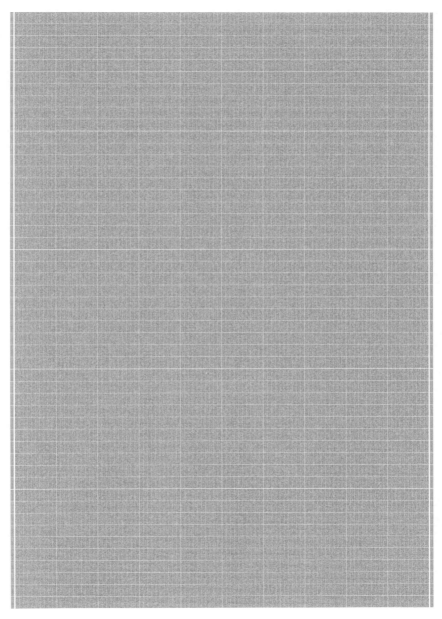

π upto 100,000,000 decimal digits　　　　62000001–62500000

円周率 100,000,000 桁表　　　63000001–63500000

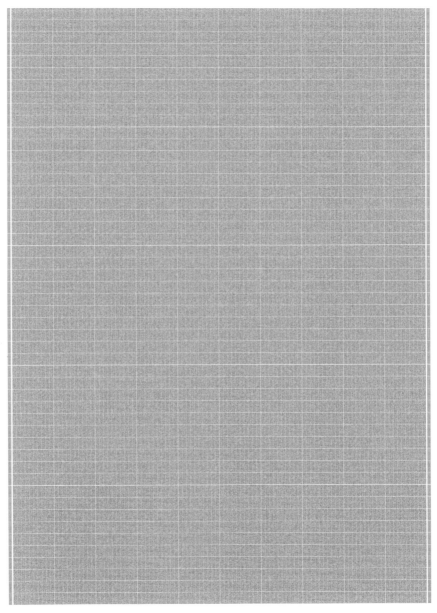

π upto 100,000,000 decimal digits　　　63000001–63500000

63500001-64000000 円周率100,000,000桁表

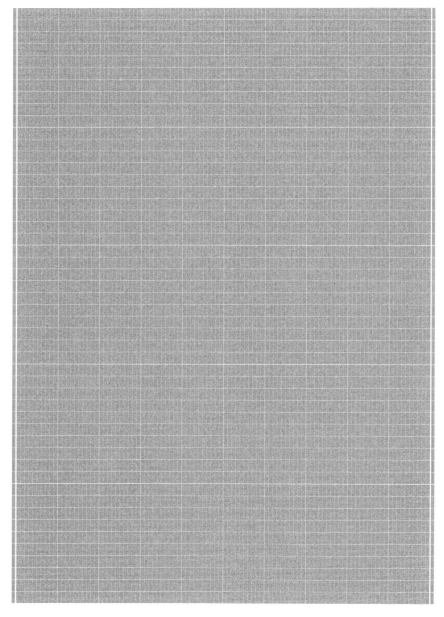

π upto 100,000,000 decimal digits

円周率 100,000,000 桁表　　　　　　　64000001–64500000

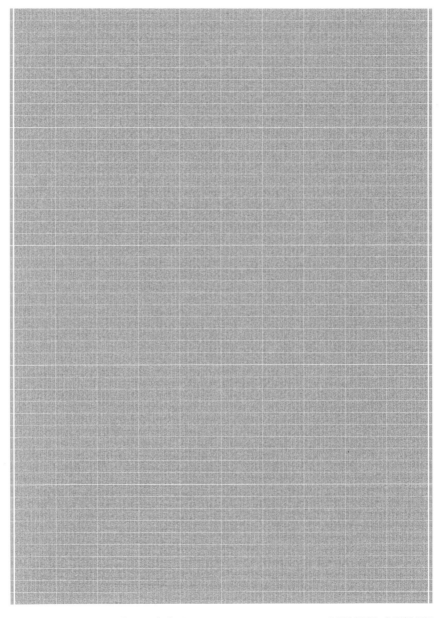

π upto 100,000,000 decimal digits　　　　64000001–64500000

64500001-65000000 円周率100,000,000桁表

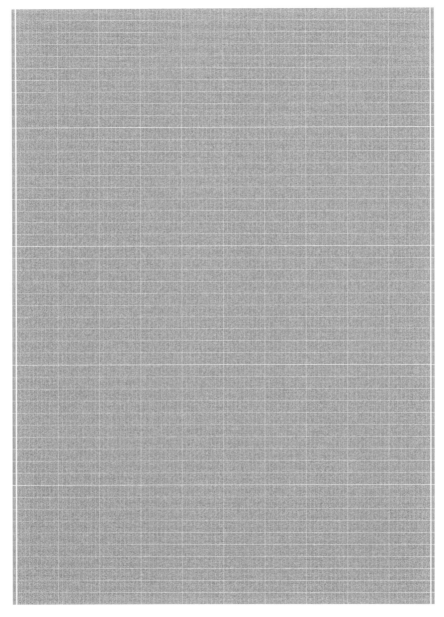

64500001-65000000 π upto 100,000,000 decimal digits

円周率 100,000,000 桁表 65000001–65500000

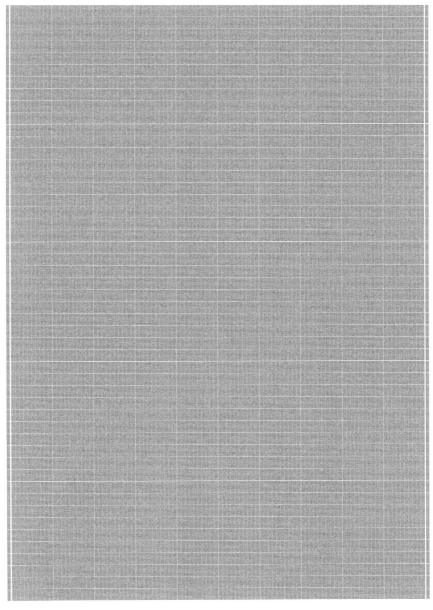

π upto 100,000,000 decimal digits 65000001–65500000

65500001–66000000 円周率 100,000,000 桁表

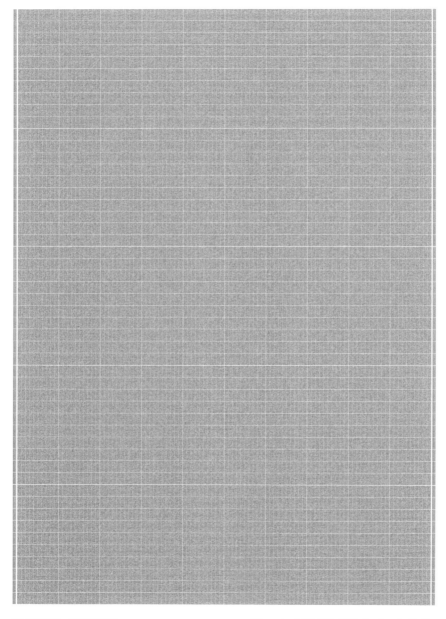

65500001–66000000 π upto 100,000,000 decimal digits

円周率 100,000,000 桁表 66000001–66500000

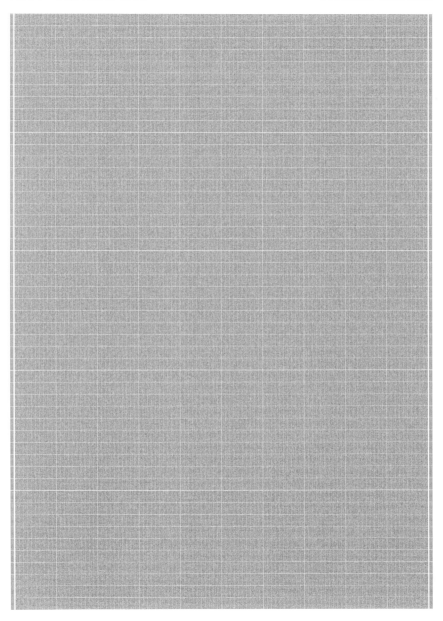

π upto 100,000,000 decimal digits 66000001–66500000

66500001-67000000 円周率 100,000,000 桁表

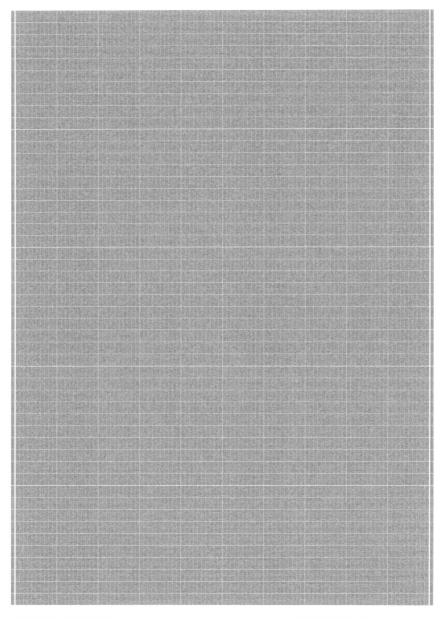

66500001-67000000 π upto 100,000,000 decimal digits

円周率 100,000,000 桁表　　　　　　　67000001–67500000

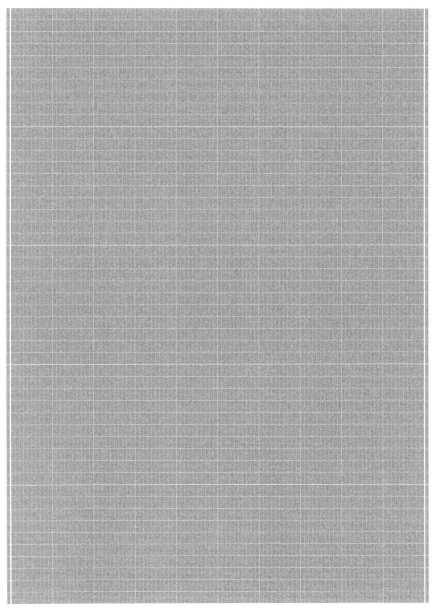

π upto 100,000,000 decimal digits　　　67000001–67500000

67500001–68000000 円周率 100,000,000 桁表

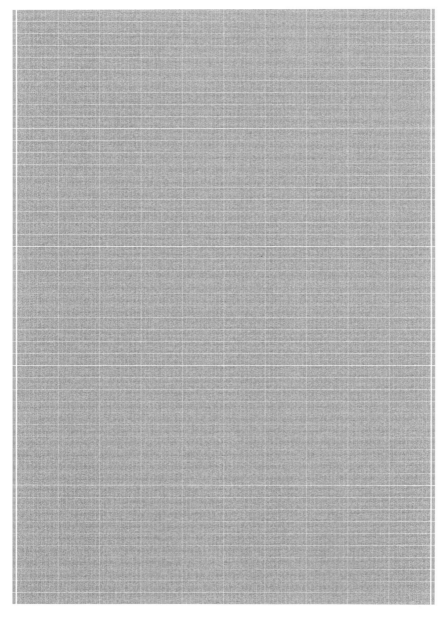

π upto 100,000,000 decimal digits

円周率 100,000,000 桁表　　　　　　　　68000001–68500000

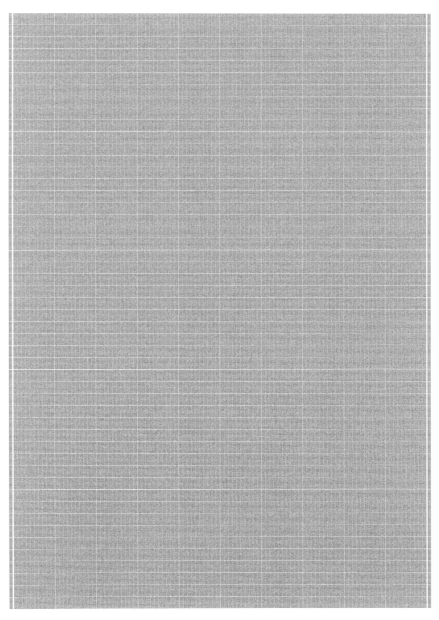

π upto 100,000,000 decimal digits　　　　　68000001–68500000

68500001–69000000 円周率 100,000,000 桁表

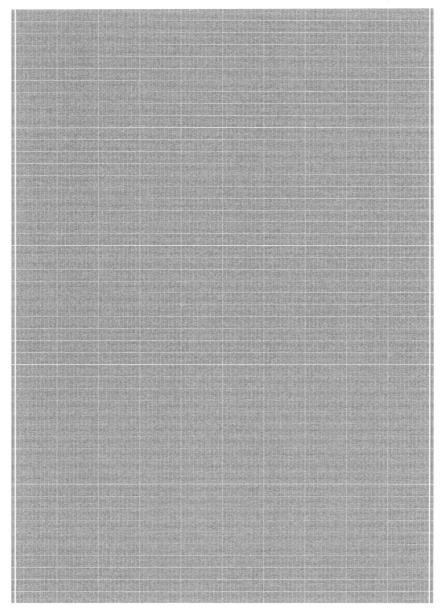

68500001–69000000 π upto 100,000,000 decimal digits

円周率 100,000,000 桁表　　　　　　69000001–69500000

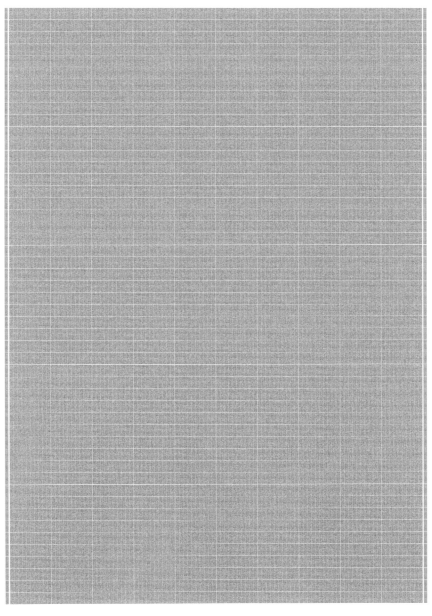

π upto 100,000,000 decimal digits　　　　69000001–69500000

69500001–70000000 円周率 100,000,000 桁表

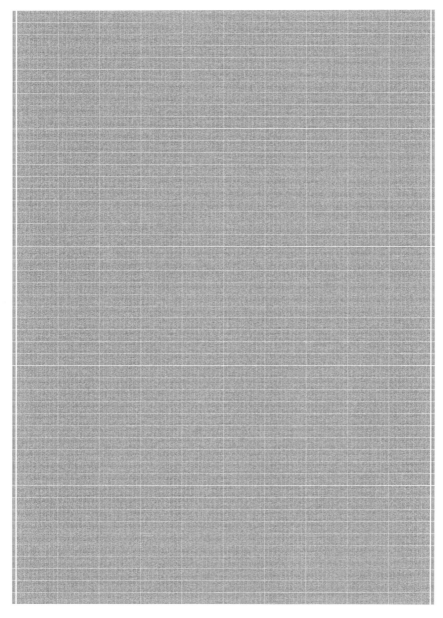

69500001–70000000 π upto 100,000,000 decimal digits

円周率 100,000,000 桁表　　　　　　　　　70000001–70500000

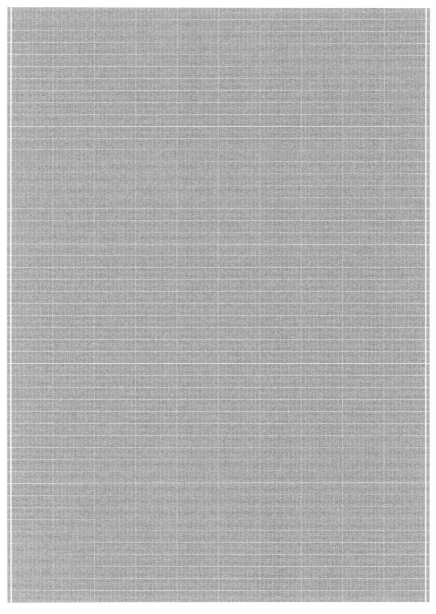

π upto 100,000,000 decimal digits　　　　　　70000001–70500000

70500001-71000000 円周率 100,000,000 桁表

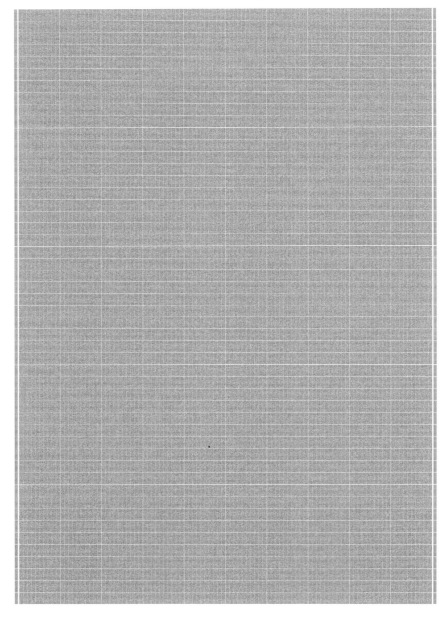

70500001-71000000 π upto 100,000,000 decimal digits

円周率 100,000,000 桁表　　　　　　　　　　　　71000001–71500000

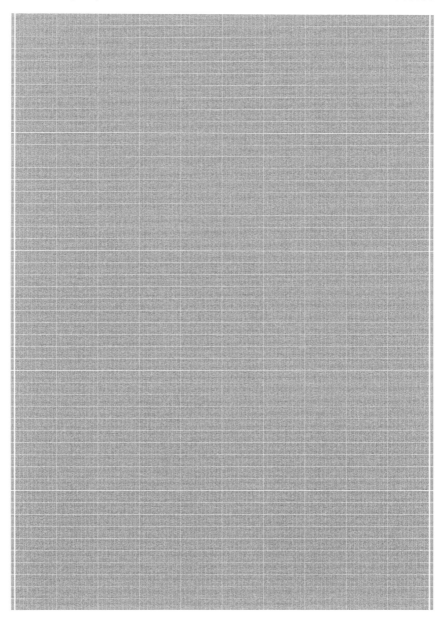

π upto 100,000,000 decimal digits　　　　　　71000001–71500000

円周率 100,000,000 桁表　　　　　　　　72000001–72500000

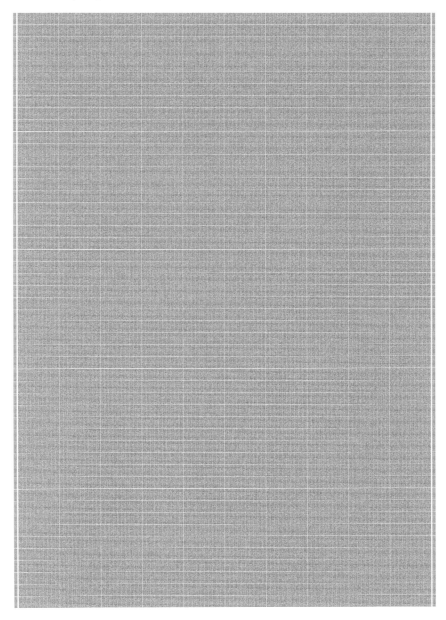

π upto 100,000,000 decimal digits　　　　　　72000001–72500000

円周率 100,000,000 桁表　　　　　　73000001–73500000

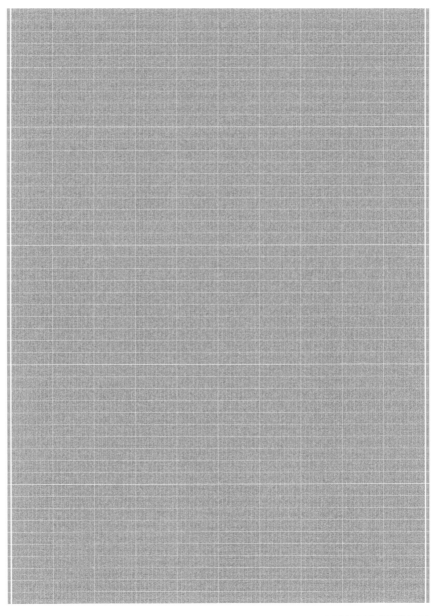

π upto 100,000,000 decimal digits　　　　73000001–73500000

73500001–74000000 円周率 100,000,000 桁表

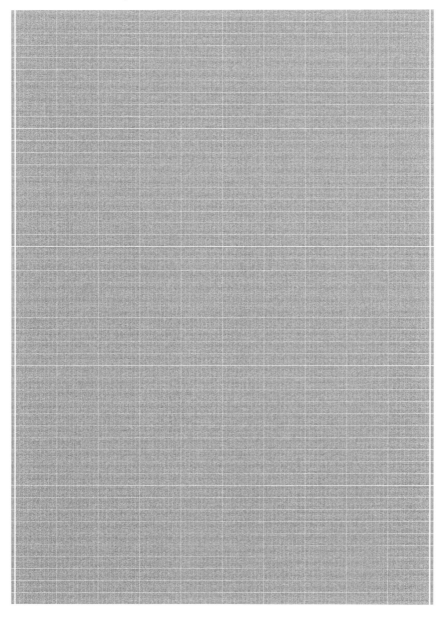

73500001–74000000 π upto 100,000,000 decimal digits

円周率 100,000,000 桁表　　　　　　　74000001–74500000

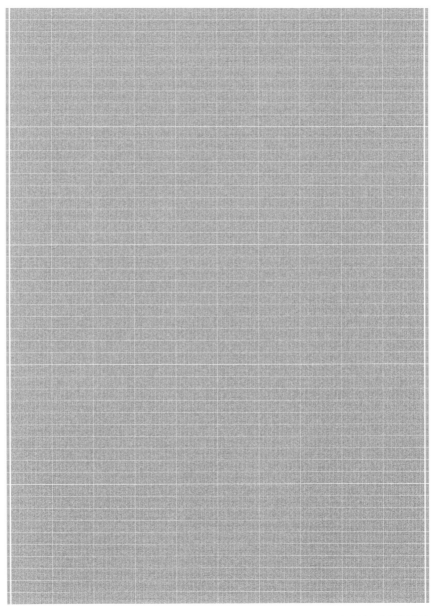

π upto 100,000,000 decimal digits　　　74000001–74500000

74500001–75000000 円周率 100,000,000 桁表

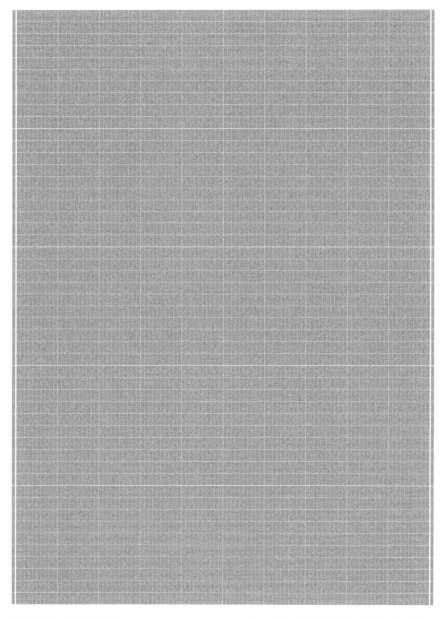

74500001–75000000 π upto 100,000,000 decimal digits

円周率 100,000,000 桁表　　　　　　　　75000001–75500000

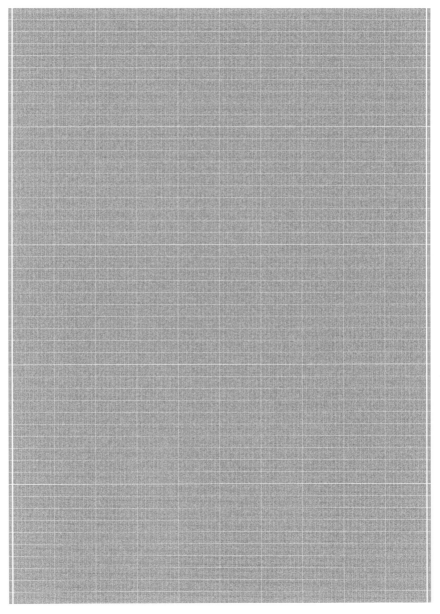

π upto 100,000,000 decimal digits　　　　　　75000001–75500000

75500001–76000000 円周率 100,000,000 桁表

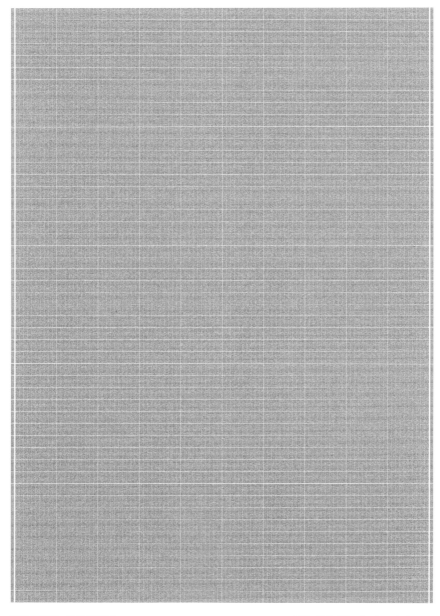

π upto 100,000,000 decimal digits

円周率 100,000,000 桁表　　　　　　　　　　76000001–76500000

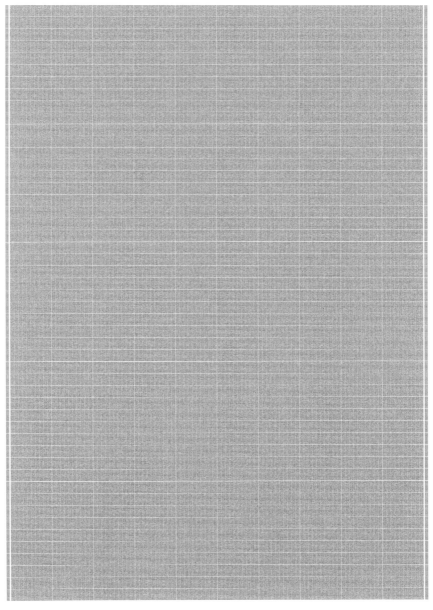

π upto 100,000,000 decimal digits　　　　76000001–76500000

76500001–77000000 円周率 100,000,000 桁表

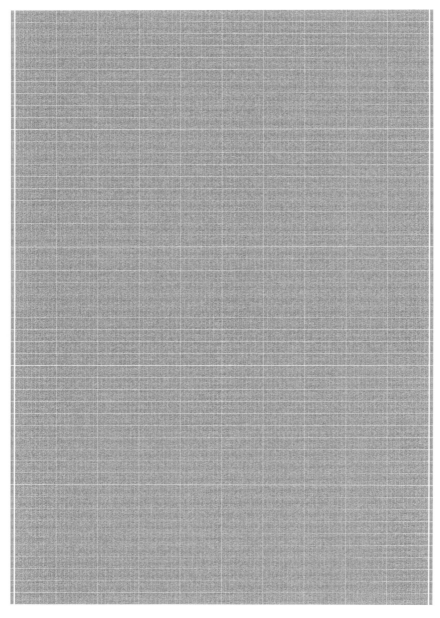

76500001–77000000 π upto 100,000,000 decimal digits

円周率 100,000,000 桁表　　　　77000001–77500000

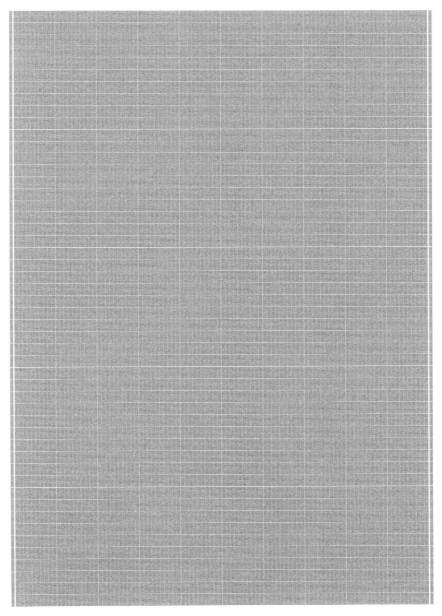

π upto 100,000,000 decimal digits　　　　77000001–77500000

77500001–78000000 円周率 100,000,000 桁表

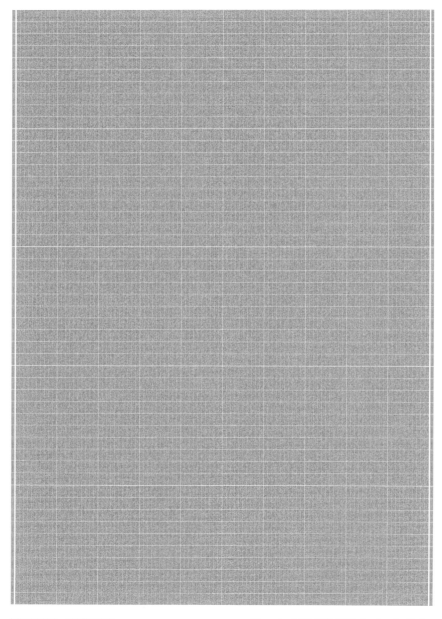

77500001–78000000 π upto 100,000,000 decimal digits

円周率 100,000,000 桁表　　　　　　　　78000001–78500000

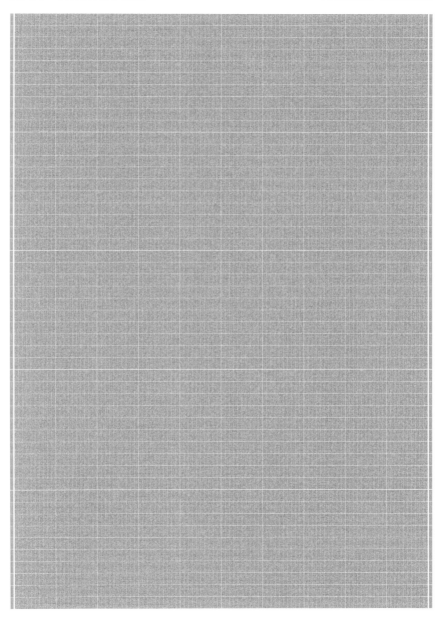

π upto 100,000,000 decimal digits　　　　78000001–78500000

78500001-79000000 円周率 100,000,000 桁表

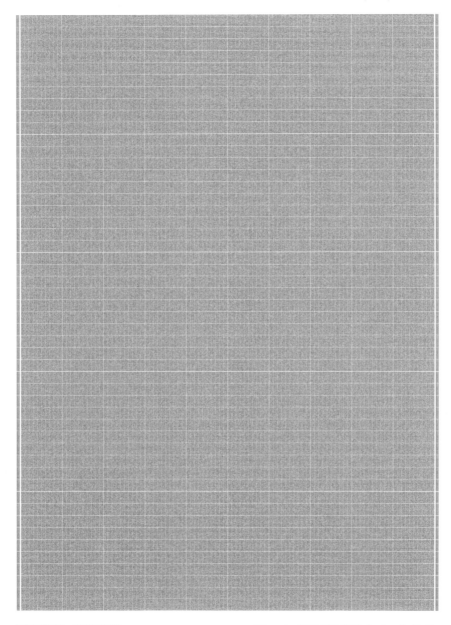

78500001-79000000 π upto 100,000,000 decimal digits

円周率 100,000,000 桁表　　　　　　　79000001–79500000

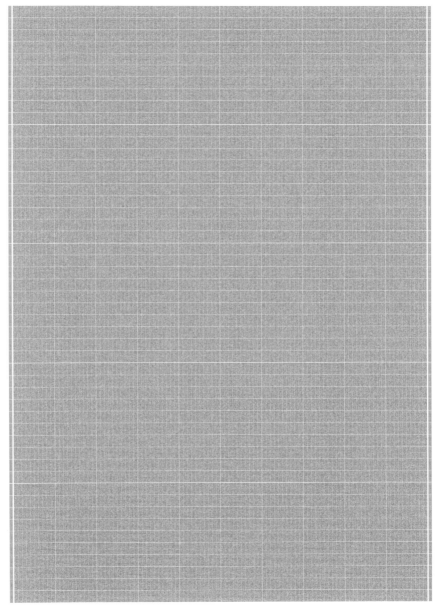

π upto 100,000,000 decimal digits　　　　　79000001–79500000

79500001-80000000 円周率 100,000,000 桁表

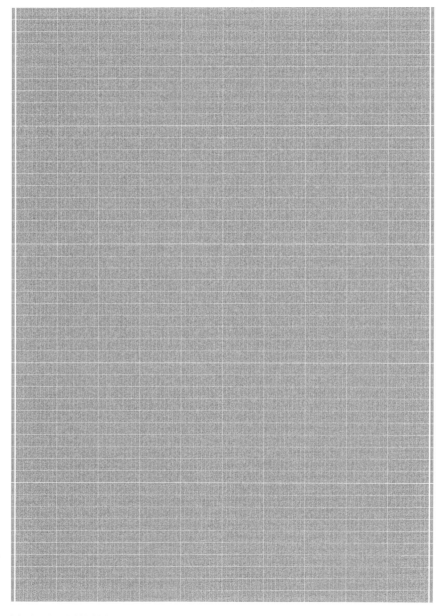

79500001-80000000 π upto 100,000,000 decimal digits

円周率 100,000,000 桁表　　　　　　　　　　　80000001–80500000

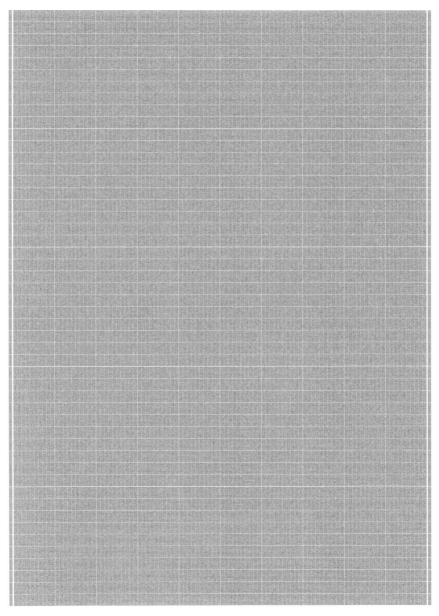

π upto 100,000,000 decimal digits　　　　　　80000001–80500000

80500001–81000000 円周率 100,000,000 桁表

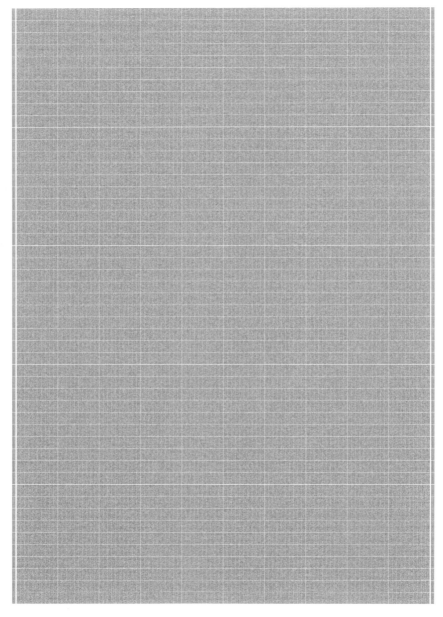

80500001–81000000 π upto 100,000,000 decimal digits

円周率 100,000,000 桁表　　　　　　　　　　81000001–81500000

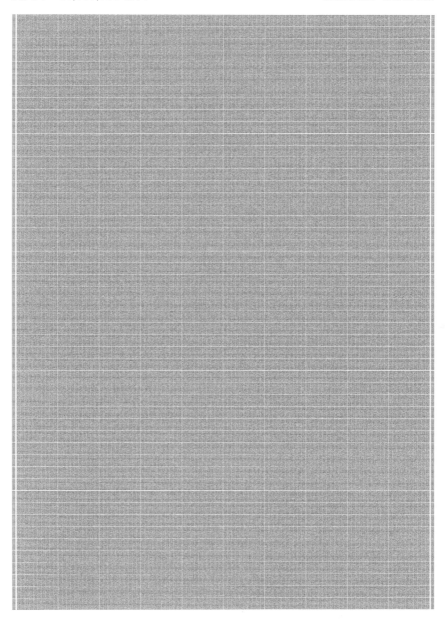

π upto 100,000,000 decimal digits　　　　　81000001–81500000

円周率 100,000,000 桁表　　　　　　　　　　82000001–82500000

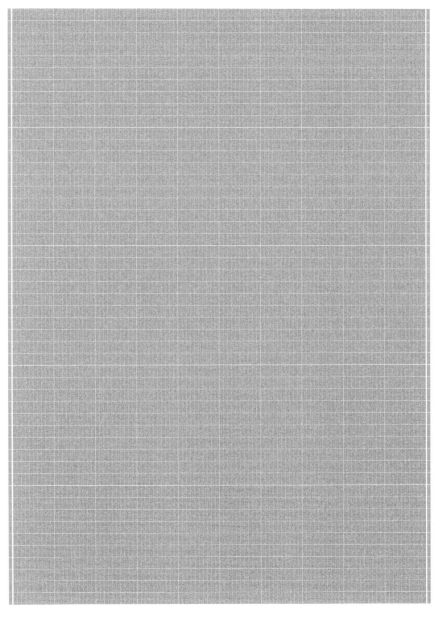

π upto 100,000,000 decimal digits　　　　　82000001–82500000

82500001-83000000 円周率 100,000,000 桁表

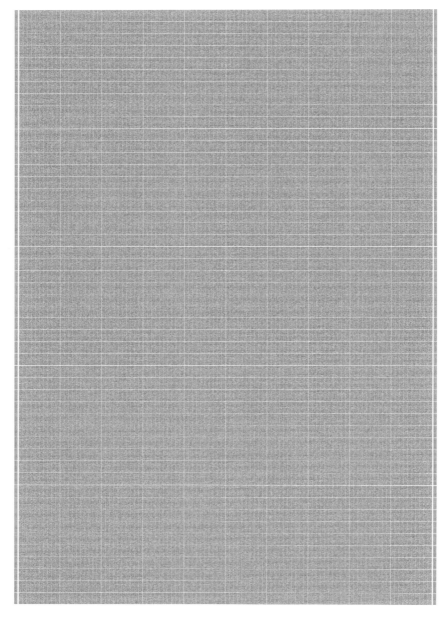

82500001-83000000 π upto 100,000,000 decimal digits

円周率 100,000,000 桁表　　　　　　　　　83000001–83500000

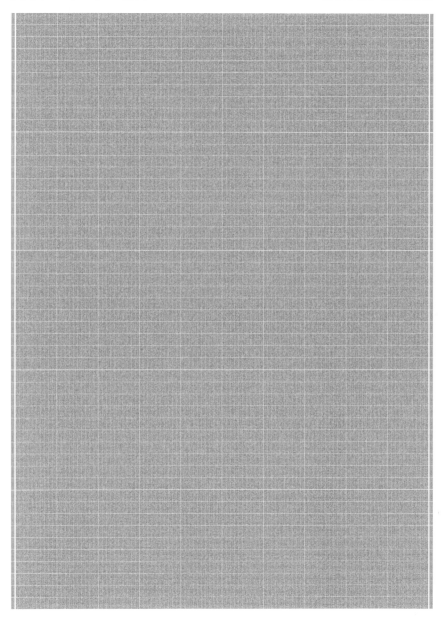

π upto 100,000,000 decimal digits　　　　83000001–83500000

83500001–84000000 円周率100,000,000桁表

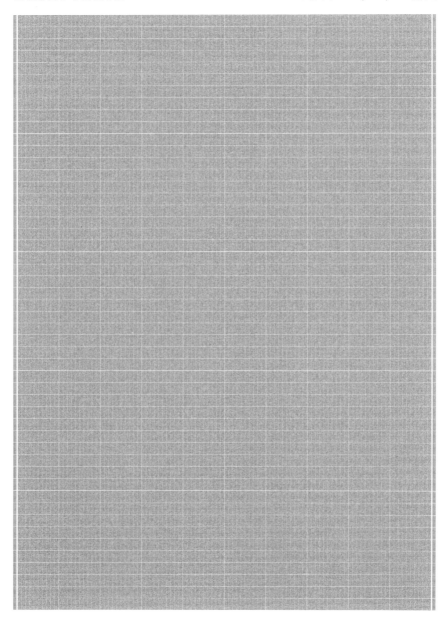

83500001–84000000 π upto 100,000,000 decimal digits

円周率 100,000,000 桁表 84000001–84500000

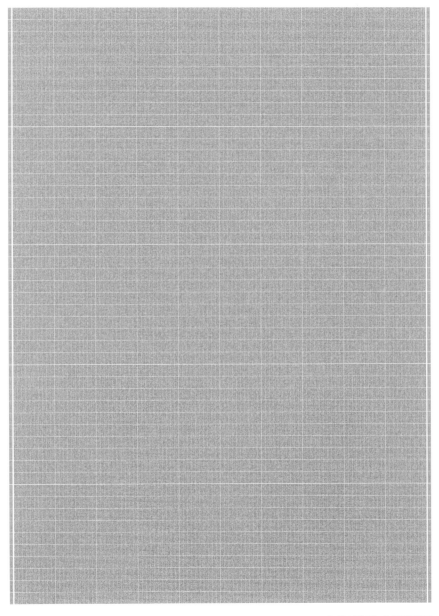

π upto 100,000,000 decimal digits 84000001–84500000

84500001–85000000 円周率 100,000,000 桁表

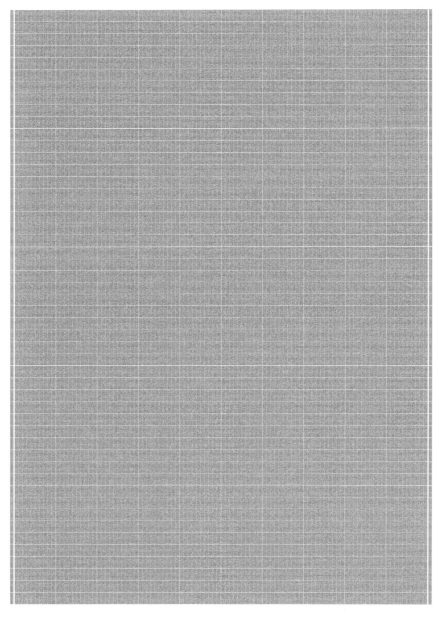

84500001–85000000 π upto 100,000,000 decimal digits

円周率 100,000,000 桁表　　　　　　　　　85000001–85500000

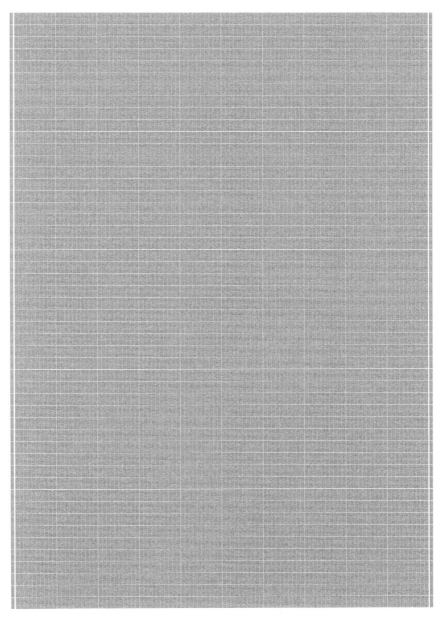

π upto 100,000,000 decimal digits　　　　　85000001–85500000

85500001-86000000 円周率 100,000,000 桁表

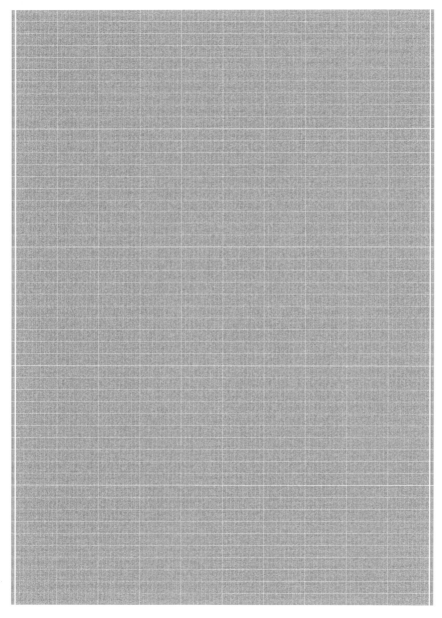

85500001-86000000 π upto 100,000,000 decimal digits

円周率 100,000,000 桁表 86000001–86500000

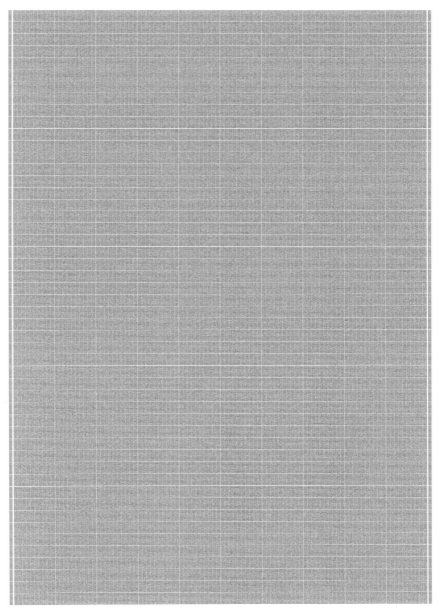

π upto 100,000,000 decimal digits 86000001–86500000

86500001-87000000 円周率 100,000,000 桁表

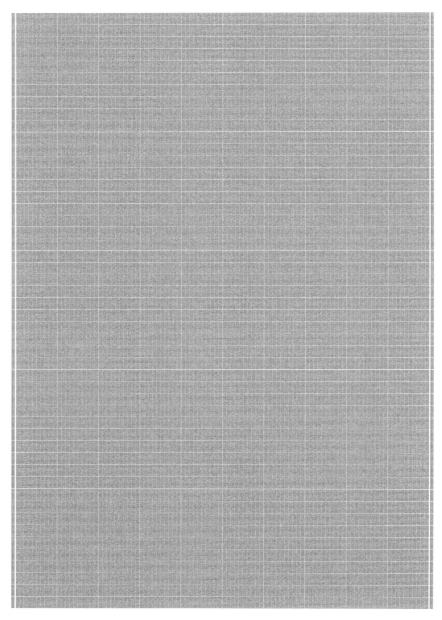

86500001-87000000 π upto 100,000,000 decimal digits

円周率 100,000,000 桁表　　　　　　　　　　87000001–87500000

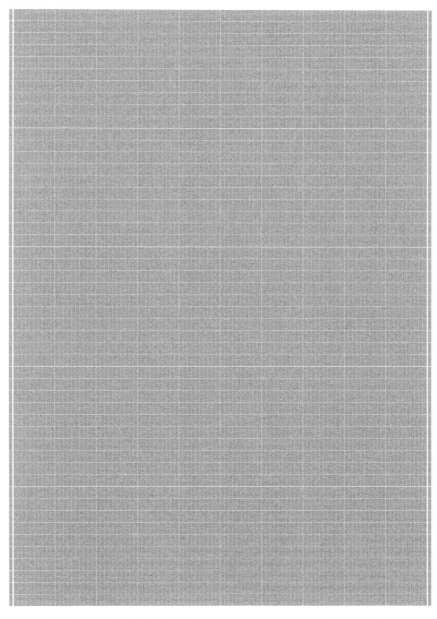

π upto 100,000,000 decimal digits　　　　　　87000001–87500000

87500001–88000000 円周率 100,000,000 桁表

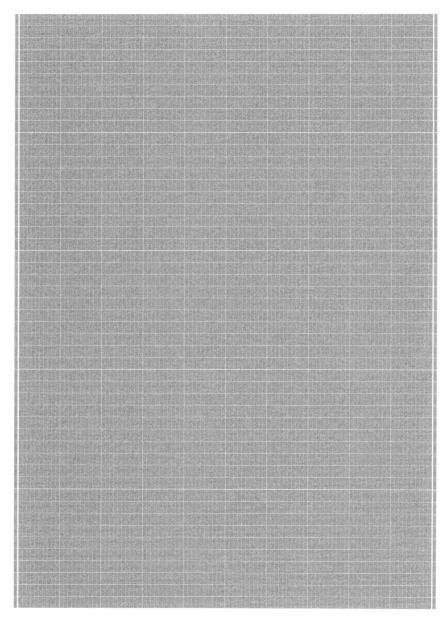

87500001–88000000 π upto 100,000,000 decimal digits

円周率 100,000,000 桁表　　　　　　　　88000001–88500000

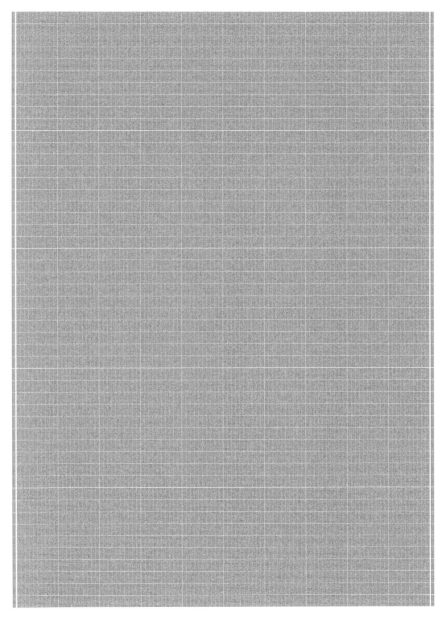

π upto 100,000,000 decimal digits　　　　88000001–88500000

88500001-89000000 円周率 100,000,000 桁表

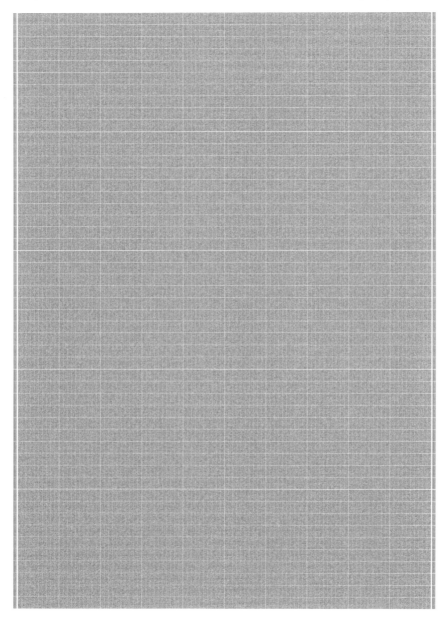

88500001-89000000 π upto 100,000,000 decimal digits

円周率 100,000,000 桁表　　　　　　　　89000001–89500000

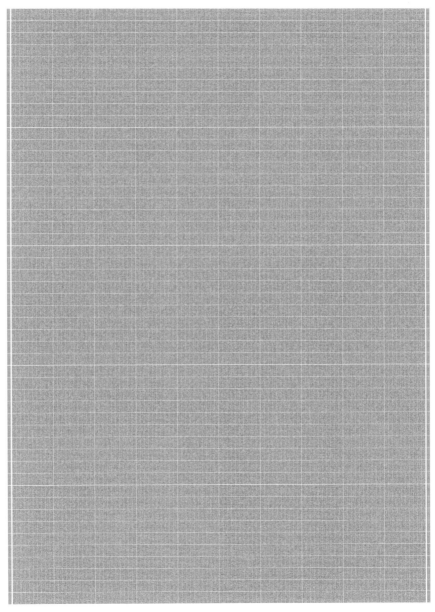

π upto 100,000,000 decimal digits　　　　89000001–89500000

円周率 100,000,000 桁表　　　　　　　　　　90000001–90500000

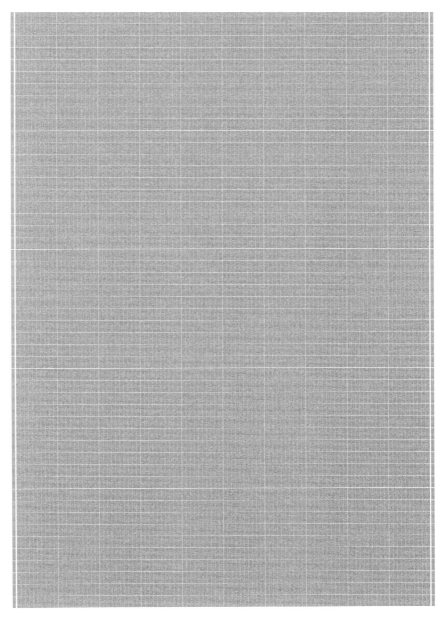

π upto 100,000,000 decimal digits　　　　90000001–90500000

90500001-91000000 円周率 100,000,000 桁表

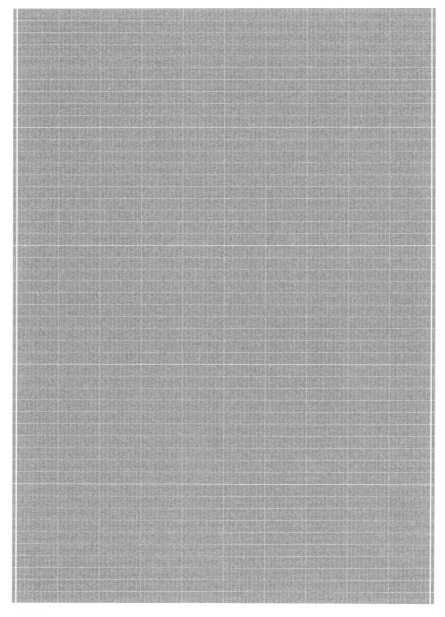

90500001-91000000 π upto 100,000,000 decimal digits

円周率 100,000,000 桁表 91000001–91500000

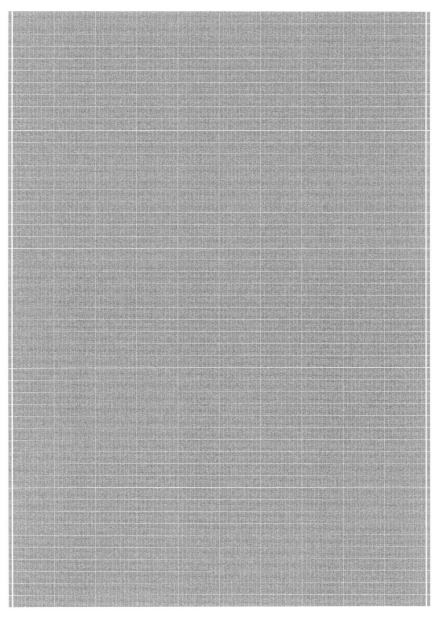

π upto 100,000,000 decimal digits 91000001–91500000

91500001-92000000 円周率 100,000,000 桁表

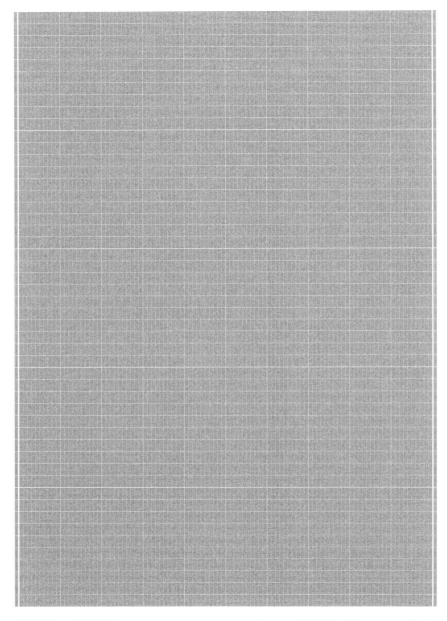

91500001-92000000 π upto 100,000,000 decimal digits

円周率 100,000,000 桁表　　　　　　　　　　92000001–92500000

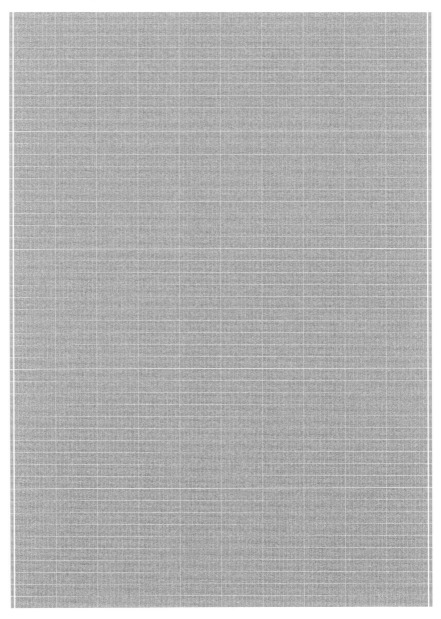

π upto 100,000,000 decimal digits　　　　　92000001–92500000

92500001–93000000 円周率 100,000,000 桁表

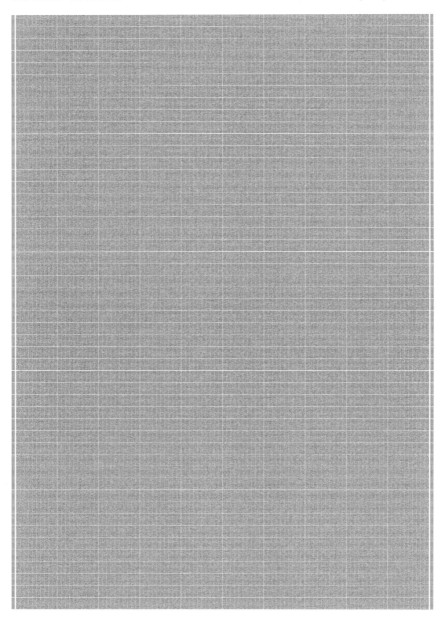

92500001–93000000 π upto 100,000,000 decimal digits

円周率 100,000,000 桁表　　　　　　　　93000001–93500000

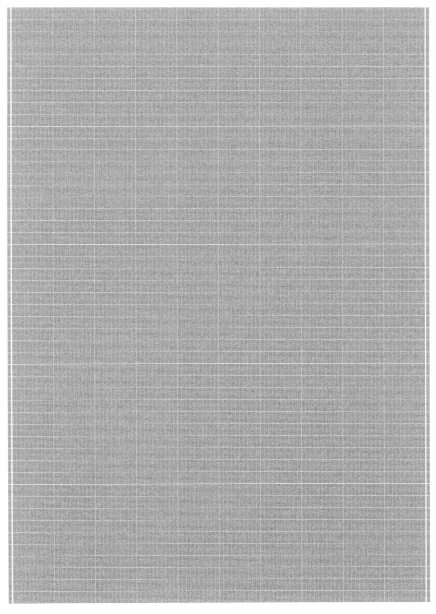

π upto 100,000,000 decimal digits　　　　93000001–93500000

93500001-94000000 円周率 100,000,000 桁表

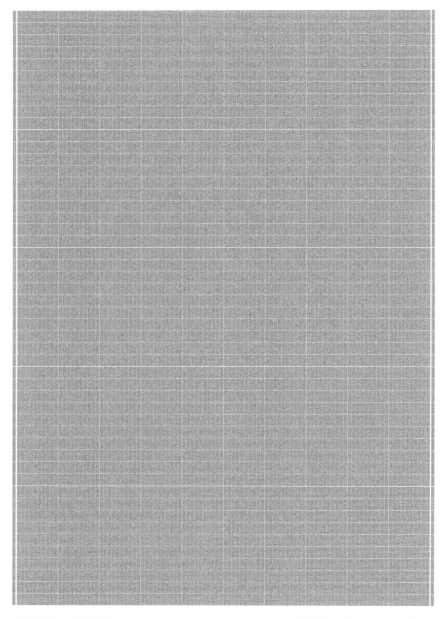

93500001-94000000 π upto 100,000,000 decimal digits

円周率 100,000,000 桁表　　　　　　　　　　94000001–94500000

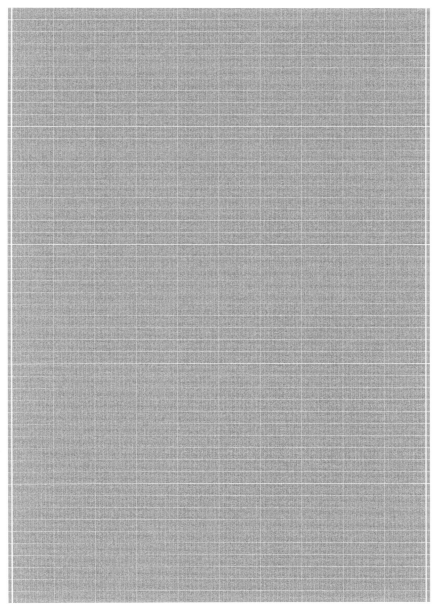

π upto 100,000,000 decimal digits　　　　　94000001–94500000

94500001–95000000 円周率 100,000,000 桁表

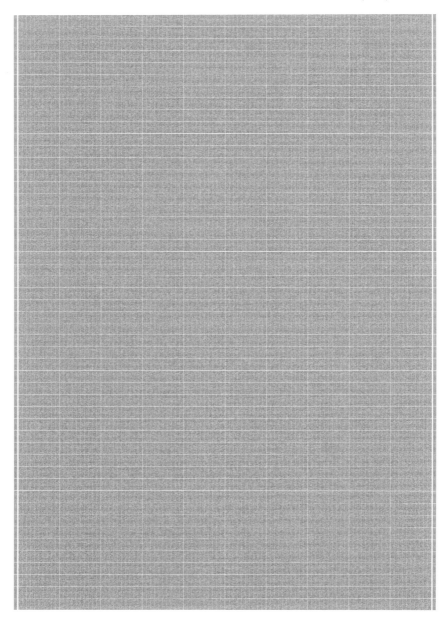

94500001–95000000 π upto 100,000,000 decimal digits

円周率 100,000,000 桁表　　　　　　　　　　　95000001–95500000

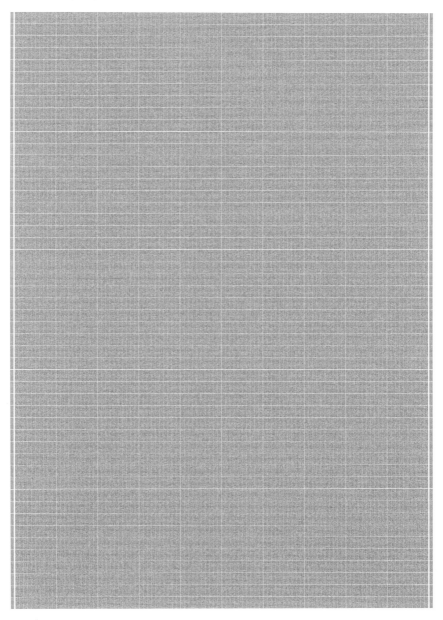

π upto 100,000,000 decimal digits　　　　　　95000001–95500000

95500001–96000000 円周率 100,000,000 桁表

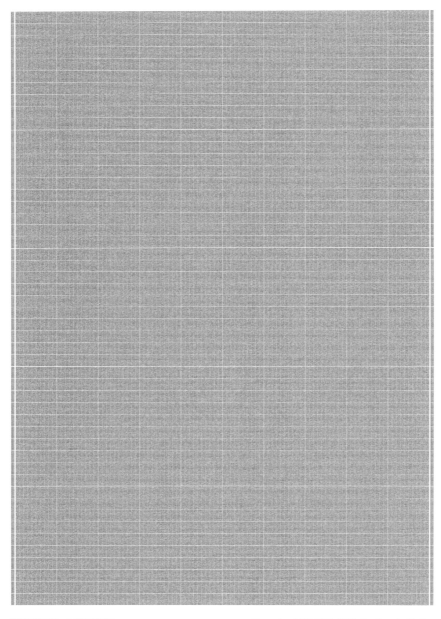

95500001–96000000 π upto 100,000,000 decimal digits

円周率 100,000,000 桁表　　　　　　　　　96000001–96500000

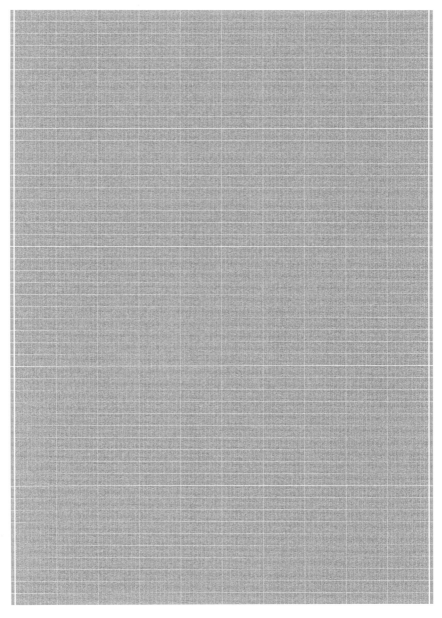

π upto 100,000,000 decimal digits　　　　96000001–96500000

96500001-97000000 円周率100,000,000桁表

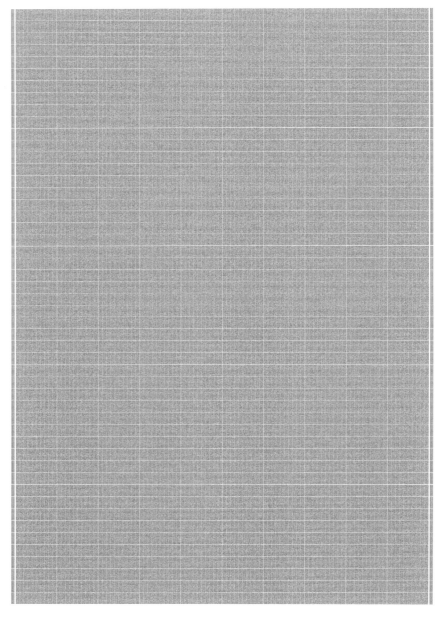

96500001-97000000 π upto 100,000,000 decimal digits

円周率 100,000,000 桁表 97000001–97500000

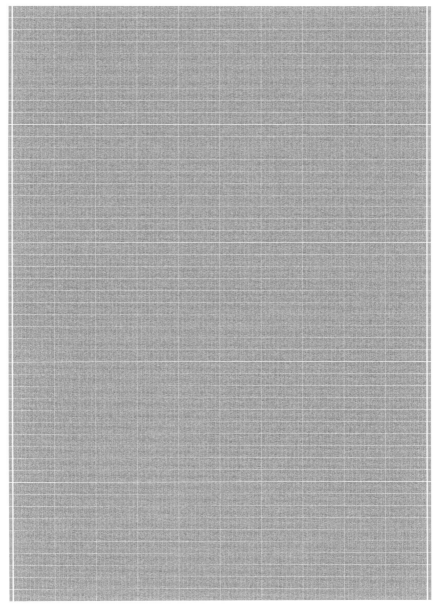

π upto 100,000,000 decimal digits 97000001–97500000

97500001–98000000 円周率 100,000,000 桁表

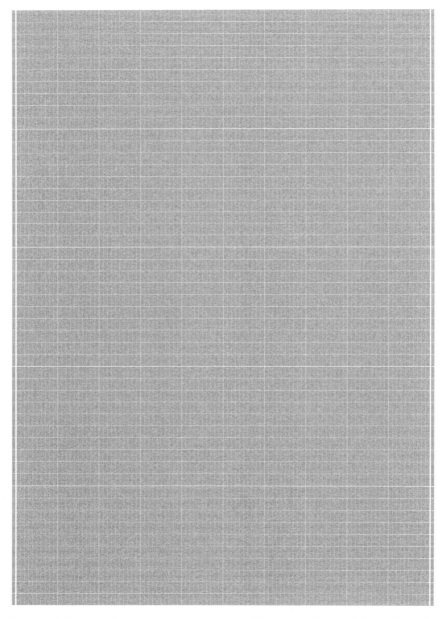

97500001–98000000 π upto 100,000,000 decimal digits

円周率 100,000,000 桁表　　　　　　　　98000001–98500000

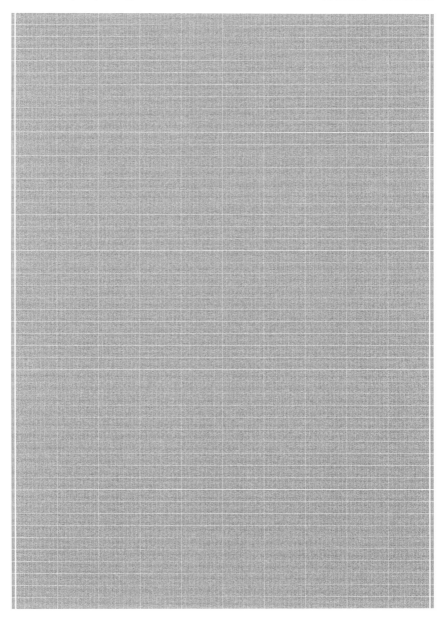

π upto 100,000,000 decimal digits　　　　98000001–98500000

98500001-99000000 円周率 100,000,000 桁表

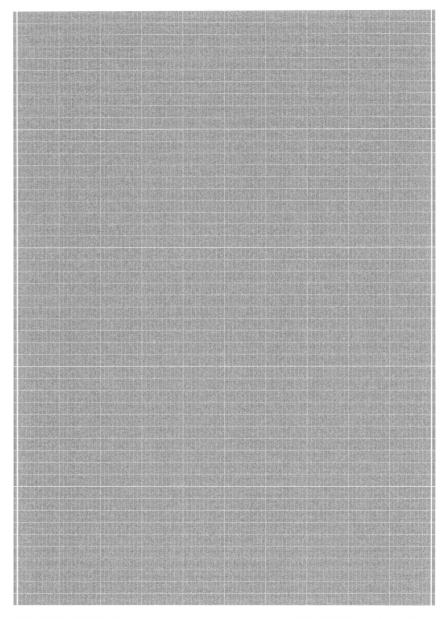

98500001-99000000 π upto 100,000,000 decimal digits

円周率 100,000,000 桁表 99000001–99500000

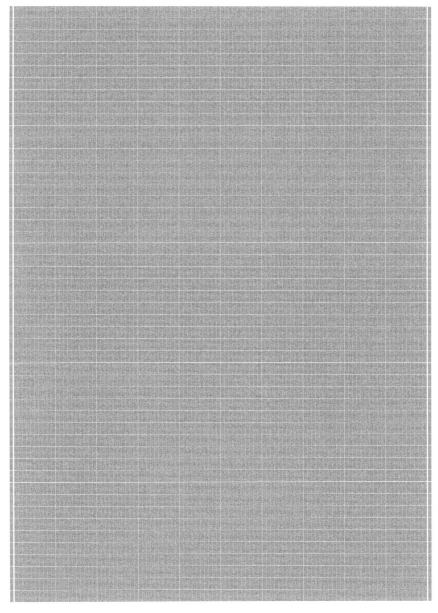

π upto 100,000,000 decimal digits 99000001–99500000

99500001–100000000 円周率 100,000,000 桁表

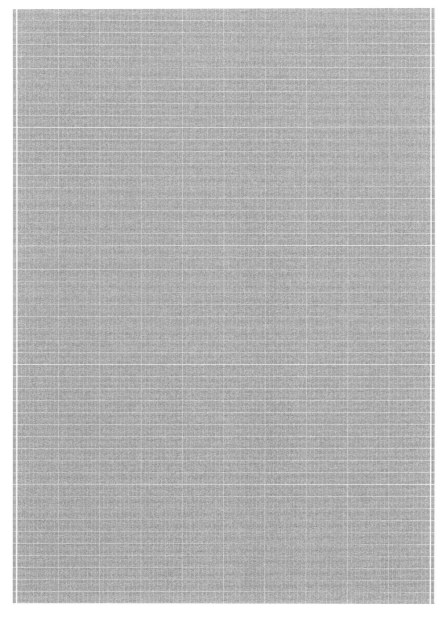

99500001–100000000 π upto 100,000,000 decimal digits

本書で使用している精細印刷用のフォント
2400 dpi　横 10 ドット　縦 30 ドット

345

678

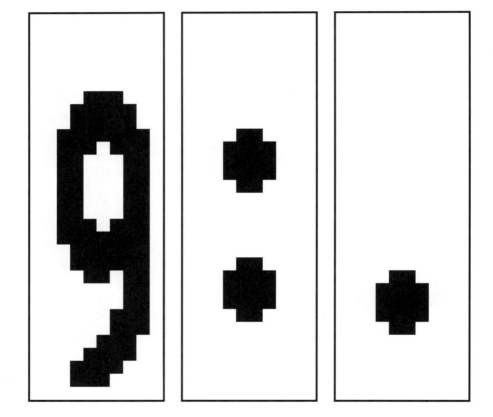

Q. なにを血迷ってこんな本を作ったんですか？

A. そんなふうに思う人はこの本を買わないと思います。

Q. これは単なるスクリーントーンじゃないんですか？

A. そんなこと言うと、これを印刷した人が泣きますよ。

Q. こんな本を売るなんて、手抜きなんじゃないんですか？

A. 我々はこの本のために、精細印刷用のフォントを作成しました。ふつうの本程度に手間はかかっていると思います。

Q. 著作権はどうなっていますか？

A. 円周率は創作物ではなく、この本はただの事実の羅列なので、この本の主要部分に著作権はありません。他の部分についても著作権を放棄します。引用・転載・複製など自由にやっていただいてけっこうです。

Q. 円周率はこれで全部ですか？

A. まさか。無限に続きます。この本 100,000,000 冊でも足りません。

円周率 100,000,000 桁表 縮刷版
2019 年 10 月 13 日 初版 発行
2019 年 10 月 26 日 初版 第 2 刷 発行
2021 年 10 月 13 日 初版 第 3 刷（フォント改訂）発行
2022 年 10 月 13 日 初版 第 4 刷 発行
2025 年 10 月 13 日 初版 第 5 刷 発行

著 者　真実のみを記述する会　（しんじつのみをきじゅつするかい）
発行者　星野 香奈　（ほしの かな）
発行所　同人集合 暗黒通信団　(https://ankokudan.org/d/)
　　　　〒277-8691 千葉県柏局私書箱 54 号 D 係
印刷所　有限会社 ねこのしっぽ
　　　　製版機　三菱製紙株式会社 FREDIA Eco Wz
　　　　印刷機　株式会社小森コーポレーション SPICA 26P
　　　　本文紙　北越コーポレーション株式会社 上質紙 キンマリ SW
本 体　628 円 / ISBN978-4-87310-628-1 C3041

本書の内容の一部または全部を無断で複写複製（コピー）することは、法律で認められた場合に相当し、著作者および出版者の権利の侵害となることはないので、やりたければ勝手にやって下さい。

Ⓒ Copyleft 2019–2025　暗黒通信団　　　　　Printed in Japan